Developing the Blue Economy

Robert C. Brears

Developing the Blue Economy

palgrave
macmillan

Robert C. Brears
Our Future Water
Christchurch, New Zealand

ISBN 978-3-030-84215-4 ISBN 978-3-030-84216-1 (eBook)
https://doi.org/10.1007/978-3-030-84216-1

Cover credit: Miguel Navarro

This Palgrave Macmillan imprint is published by the registered company Springer Nature Switzerland AG
The registered company address is: Gewerbestrasse 11, 6330 Cham, Switzerland

Acknowledgements

First, I wish to thank Rachael Ballard for being a visionary who enables books like mine to come to fruition. Second, I wish to thank Mum, who has a great interest in the environment and has supported me in this journey of writing the book.

Contents

List of Tables

1

Introduction

Introduction

Traditionally, the ocean economy is viewed solely as a mechanism for economic growth. In this business-as-usual approach, large-scale industrial economies have developed the ocean economy by exploiting maritime and marine resources, often without considering how those activities impact the future health or productivity of those same resources. This has led to marine ecosystems being viewed and treated as limitless resources; the marine environment becoming a dumping ground for waste; overfishing diminishing fishing stocks; ocean habitats being degraded from coastal developments; sea-level rise impacting coastal communities and infrastructure; increasing ocean acidification; and the marginalisation of poor coastal communities.

In response, there is a transition underway worldwide towards the blue economy, which views economic development and ocean health as complementary to one another. In the blue economy, the environmental risks of and ecological degradation from economic activity are mitigated or significantly reduced. Therefore, economic activity balances

R. C. Brears, *Developing the Blue Economy*,
https://doi.org/10.1007/978-3-030-84216-1_1

the ocean ecosystems' long-term capacity to support this activity and remain healthy and resilient.[1,2]

Despite recognising the benefits of the blue economy, the concept has yet to be mainstreamed worldwide for various reasons. For instance, while our scientific knowledge of the oceans, including coastal processes, fisheries science, marine biology, and so forth, are well developed, there is a lack of understanding on how best to develop and implement technologies, investment strategies, and interdisciplinary partnerships that enhance synergies and reduce trade-offs between sectors in the development of a blue economy. There are many cases of initiatives worldwide that ignore blue economy concepts, even when they are obvious and proven to be beneficial to humans and nature.[3,4,5]

As such, there is a need for innovative policies, technologies, and financing tools to accelerate the transition towards a blue economy that is low-carbon, efficient, and clean with its growth driven by investments that reduce carbon emissions and pollution; protect and restore blue carbon ecosystems; enhance the resilience of vulnerable coastal communities; enhance food security and nutrition; and promote sustainable economic growth opportunities.[6,7]

Developing the Blue Economy examines various innovative approaches that promote cross-sectoral and multi-scalar collaboration, facilitate the integrated management of resources, foster partnerships between governments and industry, encourage research and development in new technologies in resource use and management, and scale-up investments across established and emerging sectors of the blue economy. The book also contains case studies that illustrate how locations of differing climates, lifestyles, and income levels are scaling up and mainstreaming the development of the blue economy.

The synopsis of the book is as follows:

Chapter 2: Challenges to the Traditional Ocean Economy: This chapter will first discuss the various challenges to the traditional ocean economy, including climate change, unsustainable fisheries and aquaculture, and marine pollution. The chapter will then discuss marine and coastal ecosystem degradation, biodiversity loss, degraded ecosystem services, impacts of marine energy systems, and COVID-19.

Chapter 3: The Blue Economy: This chapter will first introduce the concept of the blue economy and blue growth. Following this, the chapter will discuss the blue economy in the context of sustainable development. Finally, the chapter will review actions to guide the development of the blue economy.

Chapter 4: Sustainable Fisheries: This chapter will first introduce the concept of sustainable fisheries before discussing the various economic and technology tools available to rebuild global fish stocks. The chapter will then discuss the ecosystem approach to fisheries and the concept of ecosystem-based fisheries management.

Chapter 5: Sustainable Aquaculture: This chapter will first discuss the types of aquaculture and systems and types of structures used in brackish and marine aquaculture operations before discussing the management of waste. The chapter will then discuss the concepts of sustainable aquaculture, integrated multi-trophic aquaculture, and organic aquaculture before reviewing spatial planning for aquaculture. Finally, the chapter will examine the ecosystem approach to aquaculture.

Chapter 6: Marine Biotechnology: This chapter will first introduce the concept of marine biotechnology before discussing its application in food security, human health, and environmental recovery and restoration. The chapter will then discuss the use of microalgae and macroalgae in biofuel production. Finally, the chapter will discuss the range of supplementary products derived from macroalgae.

Chapter 7: Marine Renewable Energy: This chapter will first discuss offshore wind energy and related issues before discussing wave and tidal current energy, ocean thermal energy conversion, and salinity gradient energy. The chapter will then provide an overview of the environmental impacts of marine renewable energy (MRE) before introducing the concept of the Environmental Impact Assessment and marine spatial planning for MRE.

Chapter 8: Coastal Water Resources Management: This chapter will first discuss best management practices to mitigate pollution of surface and groundwater and the ocean before discussing watershed planning to protect water quality. The chapter will then provide an overview of aquifer storage and recovery. After which, the chapter will discuss a range of mitigation measures desalination projects can adopt to avoid or minimise environmental impacts. Finally, the chapter will discuss the prevention of marine plastic pollution.

Chapter 9: Blue Carbon Ecosystems and Ecosystem-based Adaptation: This chapter will first discuss various measures to conserve and restore blue carbon ecosystems. The chapter will then discuss seaweed production as a climate mitigation solution. Finally, the chapter will discuss the concept of ecosystem-based adaptation (EbA) and specific EbA approaches in coastal and marine areas.

Chapter 10: Blue Financing: This chapter will first provide an overview of the various sources of finance to develop the blue economy before discussing the various financing tools available to facilitate this transition.

Chapter 11: Conclusions

Notes

1. Nathan J. Bennett et al., "Towards a Sustainable and Equitable Blue Economy," *Nature Sustainability* 2, no. 11 (2019).
2. Lu Wenhai et al., "Successful Blue Economy Examples with an Emphasis on International Perspectives," *Frontiers in Marine Science* 6, no. 261 (2019).
3. Michaela Garland et al., "The Blue Economy: Identifying Geographic Concepts and Sensitivities," *Geography Compass* 13, no. 7 (2019).
4. Andrew D. L. Steven, Mathew A. Vanderklift, and Narnia Bohler-Muller, "A New Narrative for the Blue Economy and Blue Carbon," *Journal of the Indian Ocean Region* 15, no. 2 (2019).
5. Wilfried Rickels et al., "Does the European Union Achieve Comprehensive Blue Growth? Progress of Eu Coastal States in the Baltic and North Sea, and the Atlantic Ocean against Sustainable Development Goal 14," *Marine Policy* 106 (2019).
6. Bennett et al., "Towards a Sustainable and Equitable Blue Economy."
7. Wenhai et al., "Blue Economy Examples."

References

Bennett, Nathan J., Andrés M. Cisneros-Montemayor, Jessica Blythe, Jennifer J. Silver, Gerald Singh, Nathan Andrews, Antonio Calò, *et al.* "Towards a Sustainable and Equitable Blue Economy." *Nature Sustainability* 2, no. 11 (2019/11/01/ 2019): 991–93.

Garland, Michaela, Stephen Axon, Marcello Graziano, John Morrissey, and C. Patrick Heidkamp. "The Blue Economy: Identifying Geographic Concepts and Sensitivities." *Geography Compass* 13, no. 7 (2019/07/01/ 2019): e12445.

Rickels, Wilfried, Christian Weigand, Patricia Grasse, Jörn Schmidt, and Rüdiger Voss. "Does the European Union Achieve Comprehensive Blue Growth? Progress of Eu Coastal States in the Baltic and North Sea, and the Atlantic Ocean against Sustainable Development Goal 14." *Marine Policy* 106 (2019/08/01/ 2019): 103515.

Steven, Andrew D. L., Mathew A. Vanderklift, and Narnia Bohler-Muller. "A New Narrative for the Blue Economy and Blue Carbon." *Journal of the Indian Ocean Region* 15, no. 2 (2019/05/04/ 2019): 123–28.

Wenhai, Lu, Caroline Cusack, Maria Baker, Wang Tao, Chen Mingbao, Kelli Paige, Zhang Xiaofan, *et al.* "Successful Blue Economy Examples with an Emphasis on International Perspectives." [In English]. *Frontiers in Marine Science* 6, no. 261 (2019/06/07/ 2019).

2

Challenges to the Traditional Ocean Economy

Introduction

Traditionally, the ocean and its ecosystems have been viewed as cost-free spaces to dispose of waste and a source of limitless resources, resulting in excessive use and, in some cases, irreversible change of marine resources and coastal areas.[1] This chapter will first discuss the various challenges to the traditional ocean economy, including climate change, unsustainable fisheries and aquaculture, and marine pollution. The chapter will then discuss marine and coastal ecosystem degradation, biodiversity loss, degraded ecosystem services, impacts of marine energy systems, and COVID-19.

Climate Change

The world's greenhouse gas emissions have increased by around 1.5% per annum for the past decade. By 2100, this trend would lead to temperature increases of nearly four degrees Celsius,[2] resulting in many impacts on the world's oceans, including the following.

R. C. Brears, *Developing the Blue Economy*,
https://doi.org/10.1007/978-3-030-84216-1_2

Rising Temperatures

Over the past few decades, oceans have warmed at average rates of > 0.1 °C per decade in the upper 75 m and 0.015 °C per decade at 700 m depth. Temperature defines the geographical distribution of many species and their response to climate change. Rising temperature will alter habitat and cause changes in abundance through local extinctions and latitudinal expansions or shifts. Also, invasive species may displace existing species. Fisheries will be affected if target species move. If average global ocean temperatures rise by two degrees Celsius in the tropics, fish catches are expected to decrease by 40–60%, potentially jeopardising food security as many people in the tropics rely on fish for their protein intake. Meanwhile, warmer coastal waters will likely impact aquaculture, such as reduced production, resulting in their potential relocation.[3,4] Ocean warming also leads to deoxygenation, a reduction in the amount of oxygen dissolved in the ocean. Between 1958 and 2015, oxygen and heat content were highly correlated, with sharp increases in deoxygenation and ocean heat content beginning in the mid-1980s. Over the past 50 years, the open ocean has lost an estimated two percent of its oxygen; open-ocean oxygen-minimum zones have expanded by an area the size of the European Union; the volume of water completely devoid of oxygen has more than quadrupled.[5,6] The impact of ocean oxygen decline includes decreased biodiversity, shifts in species distributions, displacement or reduction in fishery resources, and expanding algal blooms.[7]

Decreasing Carbon Sink

Oceans act as a carbon sink, absorbing and storing carbon dioxide as part of the carbon cycle. The ocean takes up carbon dioxide in two steps. First, the carbon dioxide dissolves in the surface water. Second, the ocean's overturning circulation distributes it: ocean currents and mixing processes transport the dissolved carbon dioxide from the surface deep into the ocean's interior, where it accumulates over time. Over the period from 1994 to 2007, the ocean has taken up from the atmosphere around 34 gigatonnes of human-made carbon dioxide, corresponding to 31% of all anthropogenic carbon dioxide emitted during that time. The percentage of carbon dioxide taken up by the oceans has remained relatively stable over the past 200 years, but the absolute quantity has increased substantially. However, warmer ocean temperatures limit how much organic carbon is being transported into the deep ocean, creating a positive feedback loop in which carbon storage in the oceans will reduce as global temperatures rise further.[8,9]

The amount of carbon that has been removed and stored in the deep ocean has decreased by 1.5% as temperatures have risen globally. Each year, around 50 billion tonnes of new plankton flourish in the surface ocean, while six billion tonnes of dead plankton sink to deeper waters. Therefore, this 1.5% decline means about 100 million tonnes of extra plankton have remained at the surface each year, enabling carbon to return to the atmosphere quickly.[10] Phytoplankton take up carbon dioxide from the ocean as they photosynthesise. This process converts carbon dioxide into organic carbon. Some of these phytoplankton will sink into the deeper ocean or be consumed by other organisms, which sink themselves when they die. However, in warmer water, the phytoplankton and organisms are more likely to dissolve in the upper ocean before they can sink, and as they dissolve, the carbon dissolves with them. Because this carbon stays in the surface ocean, it is more easily released back into the atmosphere.[11,12] Also, the carbon dioxide dissolved in the ocean acidifies the water.[13,14]

Ocean Acidification

As oceans absorb more carbon dioxide, chemical reactions reduce seawater pH, carbonate ion concentration, and saturation states of biologically important calcium carbonate minerals. These chemical reactions are termed ocean acidification. Calcium carbonate minerals are the building blocks for the skeletons and shells of many marine organisms. In its normal state, seawater is supersaturated with regard to calcium carbonate minerals. However, continued ocean acidification results in many parts of the ocean becoming undersaturated with these minerals, affecting some organisms' ability to produce and maintain their shells. Since the Industrial Revolution, the pH of surface ocean waters has fallen by 0.1 pH units, approximately a 30% increase in acidity. It is projected that under business-as-usual emission scenarios, the ocean's surface waters could have acidity levels nearly 150% higher by the end of this century.[15]

Impacts of Climate Change on Corals and Marine Fishes

Coral reefs harbour the highest biodiversity of any ecosystem globally. Despite only covering less than 0.1% of the ocean floor, coral reefs host more than one-quarter of all marine fish species.[16] Coral reefs are vanishing with climate change, with many reefs worldwide having lost over 50% of their coral cover over the past 30–40 years, with some having lost more than 90%.[17] Increasing ocean acidification is significantly reducing the ability of reef-building corals to produce their skeletons.[18] Acidification limits coral growth by corroding pre-existing coral skeletons while simultaneously slowing the growth of new ones. The weaker reefs will be more vulnerable to erosion from storm waves and animals that drill into or eat coral.[19] Furthermore, many corals are unable to acclimatise to ocean acidification, affecting the future ecological function of reefs.[20]

Corals have a symbiotic relationship with microscopic algae called zooxanthellae which live in coral tissue. The algae use photosynthesis to produce nutrients, many of which pass to the corals' cells. The corals, in turn, emit waste products in the form of ammonium, which the algae consume as a nutrient.[21] These algae are the coral's primary food source and give them their colour. When the symbiotic relationship becomes stressed due to increased ocean temperature or pollution, the coral expels the algae from its tissue, causing it to turn very pale or white. Without the algae, the coral loses its major source of food and is more susceptible to disease.[22] In 2016, heat stress encompassed 51% of coral reefs globally. The first mass bleaching of the northern and far-northern Great

Barrier Reef killed 29% of the reef's shallow-water corals, and bleaching in the western Indian Ocean, killing 50% of coral reefs in Seychelles.[23]

Marine fishes are impacted by ocean acidification. Because the surrounding water has a lower pH, a fish's cell often comes into balance with the seawater by taking in carbonic acid. This changes the pH of the fish's blood, a condition called acidosis. Fish are sensitive to pH changes with even a slight lowering of pH in surrounding seawater resulting in additional energy required to excrete the excess acid out of its blood through its gills, kidneys, and intestines. This reduces the amount of energy available to digest food, escape predators, and reproduce. It can even slow fish growth.[24]

Sea-Level Rise and Risks to Coastal Zones

Global mean sea level has risen 21–24 cm since 1880, with about a third of that coming in just the last 25 years. The rising water level is mainly due to a combination of thermal expansion of seawater as it warms and glacial melt. In 2018, the global mean sea level was 8.1 cm above the 1993 average, the highest annual average in the satellite record (1993 to present). The global mean water level in the ocean rose by 3.6 mm per year from 2006 to 2015, which was 2.5 times the average rate of 1.4 mm per year throughout most of the twentieth century. By 2100, the global mean sea level is likely to rise at least 0.3 m above 2000 levels, even if greenhouse gas emissions follow a relatively low pathway in the coming decades. Furthermore, regional differences in sea-level rise due to natural variability in the strength of winds and ocean currents influence how much and where the deeper layers of the ocean store heat. Overall, climate change is likely to result in the following:

- Increased frequency, duration, and extent of coastal flooding
- Coastal defences being overtopped by waves or high tides more often
- Severe storms increasing in severity, and storm surge levels rising
- Ports being impacted by more frequent coastal flooding, heavy precipitation, and heavy wave action
- Erosion of beaches

- Potential for the saltwater intrusion of coastal aquifers[25,26]

It is projected that with climate change, more extreme coastal flooding events could damage assets worth up to $14.2 trillion by 2100 as rising seas inundate coastal homes and infrastructure. By 2100, without global investment in flood defences or a decrease in global emissions, the land area flooded will increase by 48%, the population impacted will increase by 52% (around 287 million people), and the infrastructure affected will increase by 46%, which comprises around 20% of global gross domestic product (GDP).[27]

Unsustainable Fishing

In 2018, global fish production was estimated to be 179 million tonnes, of which 82 million tonnes came from aquaculture production. Of the overall total, 156 million tonnes were used for human consumption and the remaining amount was used for non-food uses, such as fishmeal and fish oil. Between 1961 and 2017, global food fish consumption has increased at an annual rate of 3.1%, which is almost twice that of annual population growth (1.6%) over the same period as well as higher than that of all other animal protein foods (meat, dairy, milk, etc.), which increased by 2.1% per annum. Meanwhile, per capita fish consumption between 1961 and 2018 increased by around 1.5% per year, going from 9.0 kg in 1961 to 20.5% in 2018. Fish consumption rates differ between countries of varying levels of development: In developed countries, fish consumption has increased from 17.4 kg per capita in 1961 to 26.4 kg in 2007, before declining to 24.4 kg in 2017. In developing countries, fish consumption has increased from 5.2 kg per capita in 1961 to 19.4 kg in 2017 (an average annual rate of 2.4%). The least developed countries have increased their consumption from 6.1 kg in 1961 to 12.6 kg in 2017 (an average annual rate of 1.3%).[28] Global capture fisheries production reached a record 96.4 million tonnes in 2018, an increase of 5.4% from the average of the previous three years. The increase was attributed mainly to marine capture fisheries, with production increasing from 81.2 million tonnes in 2017 to 84.4 million tonnes in 2018.

The state of marine fishery resources has continued to decline, with the proportion of fish stocks within biologically sustainable levels having decreased from 90% in 1974 to 65.8% in 2015. The percentage of stocks fished at biologically unsustainable levels has increased from 10% in 1974 to 34.2% in 2017. Overfishing, which is the removal of a species of fish from a body of water at a rate that the species cannot replenish, negatively impacts biodiversity and ecosystem functioning and reduces fish production. It leads to negative social and economic consequences, including diminished food security and coastal communities' economic losses.[29]

Illegal, unreported, and unregulated (IUU) fishing is threatening fisheries' sustainability. IUU fishing, which involves a broad range of activities summarised in Table 2.1, is estimated to affect one in every five fish caught. Globally, gross revenues associated with IUU is around $17 billion, while the estimated loss in annual economic impact due to the diversion of this fish from the legitimate trade system is estimated to be up to $50 billion, while losses to countries' tax revenues are around $4 billion per annum. In addition to IUU fishing negatively impacting livelihoods, fish stocks, and the environment, it can also be connected to other illicit activities such as trafficking of narcotics and weapons, human trafficking, labour abuses, and even slavery.[30,31,32,33]

Global fisheries bycatch—the capture of non-target fish and ocean wildlife, including what is brought to port and what is discarded at sea, dead or dying—is a threat to the world's oceans that causes both direct and indirect ecological effects. It directly reduces megafauna populations of marine mammals, seabirds, and marine turtles—in some cases to the verge of extinction—and has extensive cascading effects on lower trophic levels.[34,35] It is estimated that 38.5 million tonnes of fish, around 40% of the total estimated annual global marine catch, can be considered bycatch.[36]

Ecological Effects of Aquaculture

Since 1970, aquaculture production has grown by 7.5% per year, reaching 82.1 million tonnes in 2018.[37] Aquaculture, or fish farming, is

Table 2.1 Illegal, Unreported, and Unregulated Fishing Activities

Activity	Description
Illegal fishing	• Conducted by national or foreign vessels in waters under the jurisdiction of a State, without the permission of that state, or in contravention of its laws and regulations; • Conducted by vessels flying the flag of States that are parties to a relevant regional fisheries management organisation but operate in contravention of the conservation and management measures adopted by that organisation and by which the States are bound, or relevant provisions of the applicable international law; or • In violation of national laws or international obligations, including those undertaken by cooperating States to a relevant regional fisheries management organisation
Unreported fishing	• Fishing that has not been reported, or have been misreported, to the relevant national authority, in contravention of national laws and regulations; or • Are undertaken in the area of competence of a relevant regional fisheries management organisation which have not been reported or have been misreported, in contravention of the reporting procedures of that organisation
Unregulated fishing	• In the area of application of a relevant regional fisheries management organisation that are conducted by vessels without nationality, or by those flying the flag of a State not party to that organisation, or by a fishing entity, in a manner that is not consistent with or contravenes the conservation and management measures of that organisation; or • In areas or for fish stocks in relation to which there are no applicable conservation or management measures and where such fishing activities are conducted in a manner inconsistent with State responsibilities for the conservation of living marine resources under international law

the controlled cultivation of freshwater and saltwater animals or plants. The sector is diverse, ranging from smallholder ponds to large-scale commercial operations. Aquaculture production is projected to reach 109 million tonnes in 2030, an increase of 32% over 2018. Aquaculture production is one of the fastest-growing animal food-producing sectors, with total production expected to exceed beef, pork, or poultry. Aquaculture contributes to food security, with aquaculture providing high-quality animal protein and other nutrients that are especially valuable for nutritionally vulnerable groups. Meanwhile, the sector provides significant economic opportunities, with an estimated 20.5 million people employed in aquaculture in 2018 alone. With capture fisheries becoming unsustainable due to overfishing, aquaculture is projected to overtake captured fisheries in supplying the world's protein requirements in the future. Already, the intensification of aquaculture has led to an increase in waste generation from the production systems. Aquaculture is like any other industry, with waste occurring as either unused inputs or by-products. A study in Japan found that one tonne of fish generated 0.8 kg of nitrogen and 0.1 kg of phosphorous on average, equivalent to the waste generated by 73 people per day. Additionally, the pollutants discharged by 63,000 tonnes of fish produced were equivalent to the waste generated by five million people. Other ecological effects of aquaculture are listed in Table 2.2.[38,39,40,41,42,43]

Marine Pollution

Land-based sources, including nutrients from excessive fertiliser and livestock waste runoff, wastewater, and industrial emissions transported to coastal waters, account for around 80% of marine pollution globally. Furthermore, land-based coastal pollution is the major source of marine plastic pollution.[44]

Nutrients and Eutrophication

Nutrients from agricultural production, wastewater, and industrial emissions can accelerate eutrophication in coastal waters. One of the main effects of eutrophication is that it creates dense algal blooms. These blooms limit light penetration, reduce growth, and cause die-offs in littoral zones while also harming predators' success in catching prey. High rates of photosynthesis associated with eutrophication can deplete

Table 2.2 Ecological Effects of Aquaculture

Ecological Effect	Description
Phytoplankton depletion	The extraction of phytoplankton and organic particulates can alter the composition of the phytoplankton, zooplankton, and meroplankton communities
Nutrient enrichment	The addition of nutrients to the water column through fish wastes and uneaten feed can potentially stimulate phytoplankton growth and cause changes in phytoplankton species composition
Localised organic enrichment of the seabed	The deposition of faeces and uneaten feed falling to the seafloor can lead to over-enrichment of the seabed due to the high organic content of the deposited particles
Depletion of dissolved oxygen	Depletion of dissolved oxygen can occur within and around fish farms due to the respiratory activities of the farmed fish and microbial degradation of phytoplankton or waste materials in sediments and the water column
Changes to physical attributes of water	Aquaculture farming can alter and reduce current speeds, which affects water residence times, impacting associated biological processes such as phytoplankton production and depletion
Habitat modification	Marine farm structures and their associated activities can potentially exclude or modify how particular species of marine mammals use critical or sensitive habitats, including foraging or feeding areas, resting or nursery areas, and migration routes

(continued)

Table 2.2 (continued)

Ecological Effect	Description
Entanglement	Physical interactions between aquaculture and marine mammals can lead to an increased risk of entanglement in structures, ropes, or non-biological wastes from the farm producer
Underwater noise	Underwater noise associated with regular, ongoing farm activities, including vessels, can exclude or attract marine mammals
Escapees mixing with wild species	Mixing of farmed and wild populations may result in a change of fitness, adaptability, and diversity or reduced survival of the wild population

dissolved inorganic carbon and raise pH levels to extreme levels during the day. They can also cause sedimentation of organic matter on the seabed. When these dense algal blooms die, microbial decomposition severely depletes dissolved oxygen, creating a hypoxic or anoxic "dead zone" lacking sufficient oxygen to support most organisms: Once the oxygen concentration decreases to a critical level, mobile species will leave the area while sessile organisms are forced to initiate survival behaviours or die from declining dissolved oxygen levels. Overall, deoxygenation and hypoxia in coastal waters due to land-based nutrient pollution is estimated to cover an area of about 245,000 square kilometres globally, with over 700 eutrophic and hypoxic coastal systems worldwide.[45,46,47]

Marine Plastic Pollution

Over 300 million tonnes of plastic are produced each year for a variety of uses. Land-based coastal pollution is the major source of marine plastic pollution, contributing nine million tonnes per annum. In comparison, land-based inland pollution, at-sea sources, and microplastics contribute 0.5 million tonnes, 1.75 million tonnes, and 0.95 million tonnes, respectively. Rivers are a major source of plastic pollution, carrying plastic waste from deep inland to the sea, of which around 94% accumulates on the seafloor, five percent ends up on beaches, and one percent on the ocean surface. The impacts of marine plastic pollution on marine ecosystems include blockage of intestinal tracts, entanglement, inflammation, oxidative stress, hormone disruption, reproduction impact, and metabolic and behavioural change. Chemical additives in plastic products, such as flame retardants, phthalates, and phenols, can leak into marine environments, causing endocrine disruption, development disorders, and reproductive abnormalities. Chemicals leached from plastics usually tend to bioaccumulate in the organisms that absorb them, and chemical concentrations are typically higher at higher trophic levels. Human biomonitoring has found compounds used for plastic production are already appearing in human blood and cells: In humans, chemicals in plastics have been associated with disease and pathologies, including endocrine disruption, cancers, development disorders, and reproductive abnormalities.[48,49,50,51,52,53,54]

Biodiversity Loss and Marine and Coastal Ecosystem Degradation

The world's oceans contain between 500,000 and 10 million marine species, with up to 2,000 new species described each year. The World Wildlife Fund's (WWF) Living Plant 2020 Report shows an average 68% reduction in population sizes of mammals, birds, fish, amphibians, and reptiles since 1970.[55] Already, an earlier WWF report found that marine vertebrates had declined by 49% between 1970 and 2012. By 2100, more than half the world's marine species will be at risk of extinction without significant change.[56,57]

Marine ecosystems are degraded whenever the ecosystem goods and services humans obtain from them are reduced. Marine ecosystem degradation occurs at the structural level, through alterations of biodiversity, and at the functional level, through alterations of ecosystem functioning, specifically, the altering of interactions between the living and non-living components of ecological systems that guarantee high levels of production at all trophic levels, from primary producers to top predators.[58] Currently, 60% of the world's major marine ecosystems are degraded or unsustainable due to ocean warming, acidification, and other anthropogenic activities.[59]

Coastal ecosystems are at risk, with around 40% of the world's population living within 100 kms of the coast, placing an unsustainable strain on resources. For example, coastal systems such as mangroves, salt marshes, and seagrass meadows can sequester carbon at rates of up to 50 times those of the same area of tropical forest. Yet, between 1980 and 2005, 35,000 square kilometres of mangroves were removed globally. With the human population projected to increase to more than nine billion by 2050, coastal areas will be placed under further strain.[60,61]

Overall, biodiversity loss and marine and coastal ecosystem degradation impair the ocean's capacity to provide a range of ecosystem services, including the following.[62,63]

Coral Reefs

Coral reefs provide some of the most productive and valuable ecosystems on Earth, with 25% of all marine species living in coral reefs. However, it is estimated that 60% of coral reefs are threatened by a combination of ocean warming, acidification, and other anthropogenic stresses, including increased fishing, coastal agriculture, deforestation, coastal development, and shipping. This percentage is predicted to rise to 90% by 2030 and 100% by 2050. The loss of coral reefs will have detrimental impacts on coastal communities and economies. Coral reefs protect coastlines from storms and erosion, provide jobs for local communities with sand and coral harvested for construction, and offer recreational opportunities. They are also a source of food and medicines. In total, around half a billion people depend on reefs for food, income, and protection. For example, the United States National Marine Fisheries Service estimates the commercial value of U.S. fisheries from coral reefs is over $100 million, and that flood damages from 100-year storm events would increase by 91% to $272 billion without coral reefs.[64,65,66,67]

Seagrass

Seagrass provides a range of ecosystem services, including stabilising the seabed, providing grazing for marine species, and providing critical habitat for commercially important fish species. Seagrass also stores vast amounts of carbon. While seagrass only occupies 0.1% of the ocean surface, it is estimated to bury 27–44 teragrams of organic carbon per year globally, accounting for 10–18% of the total carbon burial in the oceans with soil organic carbon stocks comparable to those of temperate and tropical forests, mangroves, and tidal marshes.[68] Seagrasses are one of the most rapidly declining ecosystems in the world. It is estimated that seagrasses have been disappearing at 110 square kilometres per year since 1980. The disappearance rate has accelerated from a median of 0.9% per annum before 1940 to seven percent per annum since 1990. In addition, the loss of seagrass meadows adversely impacts fishing, with seagrass supporting commercial fisheries worth as much as $3,500 per hectare per annum.[69,70]

Mangroves

Mangroves provide spawning grounds, nurseries, nutrients, and shelter for fish, reptiles, amphibians, mammals, and birds. They also provide a range of ecosystem services, including provisioning services (firewood, timber, fisheries, other forest products), regulating services (coastal protection, carbon sequestration, buffering seagrass beds from terrestrial sediment/nutrient loads), and cultural services (recreation, ecotourism, spiritual). Globally, around 20% of mangrove cover has been lost between 1980 and 2005. The primary cause of this loss is the conversion of mangrove areas for aquaculture, agriculture, infrastructure, and tourism, climate change extreme weather events, and coastal development for high human populations.[71,72] It is estimated that between 1996 and 2016, the average rate of mangrove loss was 0.21% annually, higher than the average for tropical and subtropical forest losses. At the national level, losses are recorded in 97% of the countries and territories with mangroves. Meanwhile, degradation is recorded in 76% of countries and territories with mangroves.[73]

Marine Energy Systems

Electricity generated from marine technologies increased by an estimated 13% in 2019, significantly above the levels of the past three years. However, while these systems generate renewable energy and mitigate greenhouse gas emissions, marine renewable energy (MRE) systems can impact marine animals in various ways. Turbines pose a risk of collision to marine mammals, fish, and diving seabirds. MRE systems create underwater noise, affecting marine animals, as they rely on sounds in the ocean to communicate, navigate, find food, socialise, and evade predators. Therefore, anthropogenic noise in the marine environment has the potential to interfere with these activities. Electromagnetic fields emitted by electric cables and MRE devices can potentially cause changes in behaviour, movement, and reproduction success of sedentary benthic organisms.[74,75]

Covid-19

The COVID-19 pandemic has impacted fish products and seafood value chains through falling consumer demand (restaurants and hotels), closure of retail business, disrupted trade routes, changes in consumer demands, and a potential increase in sanitary measures on fish and seafood products. The pandemic is hindering demand, fishing capacity, and output all at once. Small-scale fishers, many of whom are self-employed and do not have income or health insurance, are most at risk of being unable to sustain their families, afford operational costs, or repay outstanding loans for equipment. Other marine sectors affected by the pandemic include the travel and tourism sector, with potential GDP losses in this sector estimated to be as high as $2.1 trillion in 2020 alone. Nonetheless, COVID-19 may have, theoretically, positive effects on the sustainability of global fisheries and the restoration of the natural environment. Therefore, there is a unique opportunity to shift resources towards actions that encourage ecosystem management and support sustainable and alternative livelihoods through small-scale fishers and coastal populations' economic activities.[76]

Notes

1. World Bank, "The Potential of the Blue Economy: Increasing Long-Term Benefits of the Sustainable Use of Marine Resources for Small Island Developing States and Coastal Least Developed Countries," (2017), https://openknowledge.worldbank.org/bitstream/handle/10986/26843/115545.pdf?sequence=1&isAllowed=y.
2. UNEP, "Emissions Gap Report 2019," (2019), https://www.unenvironment.org/resources/emissions-gap-report-2019.
3. H.-O. Pörtner, D.M. Karl, P.W. Boyd,W.W.L. Cheung, S.E. Lluch-Cota, Y. Nojiri, D.N. Schmidt, and P.O. Zavialov, "Ocean Systems," *Climate Change 2014: Impacts, Adaptation, and Vulnerability. Part A: Global and Sectoral Aspects. Contribution of Working Group II to the Fifth Assessment Report of the Intergovernmental Panel on Climate Change* (2014), https://www.ipcc.ch/site/assets/uploads/2018/02/WGIIAR5-Chap6_FINAL.pdf.
4. NIWA, "Climate Change and the Oceans," https://niwa.co.nz/education-and-training/schools/students/climate-change/oceans.
5. Sunke Schmidtko, Lothar Stramma, and Martin Visbeck, "Decline in Global Oceanic Oxygen Content During the Past Five Decades," *Nature* 542 (2017).
6. Denise Breitburg et al., "Declining Oxygen in the Global Ocean and Coastal Waters," *Science* 359, no. 6371 (2018).
7. IUCN, "Ocean Deoxygenation," https://www.iucn.org/resources/issues-briefs/ocean-deoxygenation.
8. Chris M. Marsay et al., "Attenuation of Sinking Particulate Organic Carbon Flux through the Mesopelagic Ocean," *Proceedings of the National Academy of Sciences* (2015).

9. CarbonBrief, "Warming Oceans Less Able to Store Organic Carbon, Study Suggests," https://www.carbonbrief.org/warming-oceans-less-able-to-store-organic-carbon-study-suggests.
10. B. B. Cael, Kelsey Bisson, and Michael J. Follows, "How Have Recent Temperature Changes Affected the Efficiency of Ocean Biological Carbon Export?," *Limnology and Oceanography Letters* 2, no. 4 (2017).
11. Marsay et al., "Attenuation of Sinking Particulate Organic Carbon Flux through the Mesopelagic Ocean."
12. CarbonBrief, "Warming Oceans Less Able to Store Organic Carbon, Study Suggests."
13. ScienceDaily, "Ocean Sink for Human-Made Carbon Dioxide Measured."
14. Nicolas Gruber et al., "The Oceanic Sink for Anthropogenic Co2 from 1994 to 2007," *Science* 363, no. 6432 (2019).
15. NOAA, "What Is Ocean Acidification?," https://www.pmel.noaa.gov/co2/story/What+is+Ocean+Acidification%3F.
16. IUCN, "Coral Reefs and Climate Change," (2017), https://www.iucn.org/sites/dev/files/coral_reefs_and_climate_change_issues_brief_final.pdf.
17. Andrew C Baker, "Climate Change: Many Ways to Beat the Heat for Reef Corals," *Current Biology* 24, no. 24 (2014).
18. NOAA, "What Is Ocean Acidification?," https://www.pmel.noaa.gov/co2/story/What+is+Ocean+Acidification%3F.
19. Smithsonian Ocean, "Ocean Acidification," https://ocean.si.edu/ocean-life/invertebrates/ocean-acidification.
20. S. Comeau et al., "Resistance to Ocean Acidification in Coral Reef Taxa Is Not Gained by Acclimatization," *Nature Climate Change* 9, no. 6 (2019).
21. Katarzyna Frankowiak et al., "Photosymbiosis and the Expansion of Shallow-Water Corals," *Science Advances* 2, no. 11 (2016).
22. NOAA, "What Is Coral Bleaching?," https://oceanservice.noaa.gov/facts/coral_bleach.html.
23. "Fast Facts Coral Reefs," https://coast.noaa.gov/states/fast-facts/coral-reefs.html
24. Smithsonian Ocean, "Ocean Acidification."

25. Ministry for the Environment, "Adapting to Sea-Level Rise," https://www.mfe.govt.nz/climate-change/climate-change-and-government/adapting-climate-change/adapting-sea-level-rise.
26. C. Izaguirre et al., "Climate Change Risk to Global Port Operations," *Nature Climate Change* 11, no. 1 (2021).
27. Ebru Kirezci et al., "Projections of Global-Scale Extreme Sea Levels and Resulting Episodic Coastal Flooding over the 21st Century," *Scientific Reports* 10, no. 1 (2020).
28. FAO, "The State of World Fisheries and Aquaculture 2020," (2020), http://www.fao.org/documents/card/en/c/ca9229en.
29. Ibid.
30. FAO, "Growing Momentum to Close the Net on Illegal Fishing," http://www.fao.org/news/story/en/item/1137863/icode/.
31. "Illegal, Unreported and Unregulated (Iuu) Fishing," http://www.fao.org/iuu-fishing/background/what-is-iuu-fishing/en/.
32. Andrew M. Song et al., "Collateral Damage? Small-Scale Fisheries in the Global Fight against Iuu Fishing," *Fish and Fisheries* 21, no. 4 (2020).
33. U. R. Sumaila et al., "Illicit Trade in Marine Fish Catch and Its Effects on Ecosystems and People Worldwide," *Science Advances* 6, no. 9 (2020).
34. J. Senko et al., "Comparing Bycatch Mitigation Strategies for Vulnerable Marine Megafauna," *Animal Conservation* 17, no. 1 (2014).
35. Lisa M. Komoroske and Rebecca L. Lewison, "Addressing Fisheries Bycatch in a Changing World," *Frontiers in Marine Science* 2, no. 83 (2015).
36. R. W. D. Davies et al., "Defining and Estimating Global Marine Fisheries Bycatch," *Marine Policy* 33, no. 4 (2009).
37. FAO, "The State of World Fisheries and Aquaculture 2020."
38. Ibid.
39. Marcel Martinez-Porchas and Luis R. Martinez-Cordova, "World Aquaculture: Environmental Impacts and Troubleshooting Alternatives," *TheScientificWorldJournal* 2012 (2012).

40. David W. Cole et al., "Aquaculture: Environmental, Toxicological, and Health Issues," *International Journal of Hygiene and Environmental Health* 212, no. 4 (2009).
41. Ministry for Primary Industries, "Overview of Ecological Effects of Aquaculture," (2013), https://www.mpi.govt.nz/dmsdocument/4300/direct.
42. CTA Brussels Office, "Fish-Farming: The New Driver of the Blue Economy," (2013), https://europa.eu/capacity4dev/file/15656/download?token=J7_MU3_d.
43. Akeem Babatunde Dauda et al., "Waste Production in Aquaculture: Sources, Components and Managements in Different Culture Systems," *Aquaculture and Fisheries* 4, no. 3 (2019).
44. UNEP, "Land-Based Pollution," https://www.unenvironment.org/explore-topics/oceans-seas/what-we-do/working-regional-seas/land-based-pollution.
45. M. F. Chislock, Doster, E., Zitomer, R. A. & Wilson, A. E.„ "Eutrophication: Causes, Consequences, and Controls in Aquatic Ecosystems," *Nature Education Knowledge* 4, no. 4 (2013).
46. Nancy N. Rabalais et al., "Global Change and Eutrophication of Coastal Waters," *ICES Journal of Marine Science* 66, no. 7 (2009).
47. Donald F. Boesch, "Barriers and Bridges in Abating Coastal Eutrophication," *Frontiers in Marine Science* 6 (2019).
48. Bethanie Carney Almroth and Håkan Eggert, "Marine Plastic Pollution: Sources, Impacts, and Policy Issues," *Review of Environmental Economics and Policy* 13, no. 2 (2019).
49. Laurent C. M. Lebreton et al., "River Plastic Emissions to the World's Oceans," *Nature Communications* 8, no. 1 (2017).
50. J. Hammer, M. H. Kraak, and J. R. Parsons, "Plastics in the Marine Environment: The Dark Side of a Modern Gift," *Rev Environ Contam Toxicol* 220 (2012).
51. Frederic Gallo et al., "Marine Litter Plastics and Microplastics and Their Toxic Chemicals Components: The Need for Urgent Preventive Measures," *Environmental sciences Europe* 30, no. 1 (2018).
52. Martin Thiel et al., "Impacts of Marine Plastic Pollution from Continental Coasts to Subtropical Gyres—Fish, Seabirds, and Other

Vertebrates in the Se Pacific," *Frontiers in Marine Science* 5, no. 238 (2018).
53. Thomas Efferth and Norbert W. Paul, "Threats to Human Health by Great Ocean Garbage Patches," *The Lancet Planetary Health* 1, no. 8 (2017).
54. Carney Almroth and Eggert, "Marine Plastic Pollution: Sources, Impacts, and Policy Issues."
55. WWF, "Living Planet Report 2020," (2020), https://livingplanet.panda.org/en-us/.
56. "Living Blue Planet Report," (2015), http://ocean.panda.org.s3.amazonaws.com/media/Living_Blue_Planet_Report_2015_08_31.pdf.
57. UNESCO, "Facts and Figures on Marine Biodiversity," http://www.unesco.org/new/en/natural-sciences/ioc-oceans/focus-areas/rio-20-ocean/blueprint-for-the-future-we-want/marine-biodiversity/facts-and-figures-on-marine-biodiversity/.
58. ICSU-IAPSO-IUGG-SCOR, "Future of the Ocean and Its Seas: A Non-Governmental Scientific Perspective on Seven Marine Research Issues of G7 Interest," (2016), http://www.iugg.org/policy/Report_FutureOcean_G7_2016.pdf
59. UNESCO, "Facts and Figures on Marine Biodiversity."
60. UN, "Factsheet: Biodiversity," (2017), https://sustainabledevelopment.un.org/content/documents/Ocean_Factsheet_Biodiversity.pdf.
61. UNESCO, "Facts and Figures on Marine Biodiversity".
62. Ibid.
63. Boris Worm et al., "Impacts of Biodiversity Loss on Ocean Ecosystem Services," *Science* 314, no. 5800 (2006).
64. NOAA, "How Do Coral Reefs Benefit the Economy?," https://oceanservice.noaa.gov/facts/coral_economy.html
65. "Coral Reef Ecosystems," https://www.noaa.gov/education/resource-collections/marine-life/coral-reef-ecosystems#:~:text=Benefits%20of%20coral%20reef%20ecosystems,food%2C%20income%2C%20and%20protection.
66. Anna J. Woodhead et al., "Coral Reef Ecosystem Services in the Anthropocene," *Functional Ecology* 33, no. 6 (2019).

67. Michael W. Beck et al., "The Global Flood Protection Savings Provided by Coral Reefs," *Nature Communications* 9, no. 1 (2018).
68. Camila Bedulli et al., "Contribution of Seagrass Blue Carbon toward Carbon Neutral Policies in a Touristic and Environmentally-Friendly Island," *Frontiers in Marine Science* 7 (2020).
69. Michelle Waycott et al., "Accelerating Loss of Seagrasses across the Globe Threatens Coastal Ecosystems," *Proceedings of the National Academy of Sciences* (2009).
70. UN, "Factsheet: Biodiversity".
71. WWF, "Living Blue Planet Report".
72. Ilka C. Feller et al., "The State of the World's Mangroves in the 21st Century under Climate Change," *Hydrobiologia* 803, no. 1 (2017).
73. University of Cambridge IUCN, and The Nature Conservancy, "Global Assessments of Mangrove Losses and Degradation," (2018), https://www.iucn.org/sites/dev/files/content/documents/mangroveloss-brief-4pp-19.10.low_.pdf.
74. IEA, "Ocean Power," https://www.iea.org/reports/ocean-power.
75. OES-Environmental, "Oes-Environmental 2020 State of the Science Report: Environmental Effects of Marine Renewable Energy Development around the World," (2020), https://tethys.pnnl.gov/publications/state-of-the-science-2020.
76. UNCTAD, "The Covid-19 Pandemic and the Blue Economy: New Challenges and Prospects for Recovery and Resilience," (2020), https://unctad.org/en/PublicationsLibrary/ditctedinf2020d2_en.pdf.

References

Baker, Andrew C. "Climate Change: Many Ways to Beat the Heat for Reef Corals." *Current Biology* 24, no. 24 (2014/12/15/ 2014): R1166–R68.

Beck, Michael W., Iñigo J. Losada, Pelayo Menéndez, Borja G. Reguero, Pedro Díaz-Simal, and Felipe Fernández. "The Global Flood Protection Savings

Provided by Coral Reefs." *Nature Communications* 9, no. 1 (2018/06/12/ 2018): 2186.

Bedulli, Camila, Paul S. Lavery, Matt Harvey, Carlos M. Duarte, and Oscar Serrano. "Contribution of Seagrass Blue Carbon toward Carbon Neutral Policies in a Touristic and Environmentally-Friendly Island." *Frontiers in Marine Science* 7 (2020).

Boesch, Donald F. "Barriers and Bridges in Abating Coastal Eutrophication." *Frontiers in Marine Science* 6 (2019).

Breitburg, Denise, Lisa A. Levin, Andreas Oschlies, Marilaure Grégoire, Francisco P. Chavez, Daniel J. Conley, Véronique Garçon, *et al.* "Declining Oxygen in the Global Ocean and Coastal Waters." *Science* 359, no. 6371 (2018): eaam7240.

Cael, B. B., Kelsey Bisson, and Michael J. Follows. "How Have Recent Temperature Changes Affected the Efficiency of Ocean Biological Carbon Export?" *Limnology and Oceanography Letters* 2, no. 4 (2017/08/01/ 2017): 113–18.

CarbonBrief. "Warming Oceans Less Able to Store Organic Carbon, Study Suggests." https://www.carbonbrief.org/warming-oceans-less-able-to-store-organic-carbon-study-suggests.

Carney Almroth, Bethanie, and Håkan Eggert. "Marine Plastic Pollution: Sources, Impacts, and Policy Issues." *Review of Environmental Economics and Policy* 13, no. 2 (2019): 317–26.

Chislock, M. F., Doster, E., Zitomer, R. A. & Wilson, A. E.,. "Eutrophication: Causes, Consequences, and Controls in Aquatic Ecosystems." *Nature Education Knowledge* 4, no. 4 (2013): 10.

Cole, David W., Richard Cole, Steven J. Gaydos, Jon Gray, Greg Hyland, Mark L. Jacques, Nicole Powell-Dunford, Charu Sawhney, and William W. Au. "Aquaculture: Environmental, Toxicological, and Health Issues." *International Journal of Hygiene and Environmental Health* 212, no. 4 (2009/07/01/ 2009): 369–77.

Comeau, S., C. E. Cornwall, T. M. DeCarlo, S. S. Doo, R. C. Carpenter, and M. T. McCulloch. "Resistance to Ocean Acidification in Coral Reef Taxa Is Not Gained by Acclimatization." *Nature Climate Change* 9, no. 6 (2019/06/01/ 2019): 477–83.

CTA Brussels Office. "Fish-Farming: The New Driver of the Blue Economy." (2013). https://europa.eu/capacity4dev/file/15656/download?token=J7_MU3_d.

Dauda, Akeem Babatunde, Abdullateef Ajadi, Adenike Susan Tola-Fabunmi, and Ayoola Olusegun Akinwole. "Waste Production in Aquaculture:

Sources, Components and Managements in Different Culture Systems." *Aquaculture and Fisheries* 4, no. 3 (2019/05/01/ 2019): 81–88.
Davies, R. W. D., S. J. Cripps, A. Nickson, and G. Porter. "Defining and Estimating Global Marine Fisheries Bycatch." *Marine Policy* 33, no. 4 (2009): 661–72.
Efferth, Thomas, and Norbert W. Paul. "Threats to Human Health by Great Ocean Garbage Patches." *The Lancet Planetary Health* 1, no. 8 (2017): e301–e03.
FAO. "Growing Momentum to Close the Net on Illegal Fishing." http://www.fao.org/news/story/en/item/1137863/icode/.
———. "Illegal, Unreported and Unregulated (Iuu) Fishing." http://www.fao.org/iuu-fishing/background/what-is-iuu-fishing/en/.
———. "The State of World Fisheries and Aquaculture 2020." (2020). http://www.fao.org/documents/card/en/c/ca9229en.
Feller, Ilka C., Daniel A. Friess, Ken W. Krauss, and Roy R. Lewis. "The State of the World's Mangroves in the 21st Century under Climate Change." *Hydrobiologia* 803, no. 1 (2017/11/01/ 2017): 1–12.
Frankowiak, Katarzyna, Xingchen T. Wang, Daniel M. Sigman, Anne M. Gothmann, Marcelo V. Kitahara, Maciej Mazur, Anders Meibom, and Jarosław Stolarski. "Photosymbiosis and the Expansion of Shallow-Water Corals." *Science Advances* 2, no. 11 (2016): e1601122.
Gallo, Frederic, Cristina Fossi, Roland Weber, David Santillo, Joao Sousa, Imogen Ingram, Angel Nadal, and Dolores Romano. "Marine Litter Plastics and Microplastics and Their Toxic Chemicals Components: The Need for Urgent Preventive Measures." [In eng]. *Environmental sciences Europe* 30, no. 1 (2018): 13–13.
Gruber, Nicolas, Dominic Clement, Brendan R. Carter, Richard A. Feely, Steven van Heuven, Mario Hoppema, Masao Ishii, *et al.* "The Oceanic Sink for Anthropogenic Co2 from 1994 to 2007." *Science* 363, no. 6432 (2019): 1193.
Hammer, J., M. H. Kraak, and J. R. Parsons. "Plastics in the Marine Environment: The Dark Side of a Modern Gift." [In eng]. *Rev Environ Contam Toxicol* 220 (2012): 1–44.
ICSU-IAPSO-IUGG-SCOR. "Future of the Ocean and Its Seas: A Non-Governmental Scientific Perspective on Seven Marine Research Issues of G7 Interest." (2016). http://www.iugg.org/policy/Report_FutureOcean_G7_2016.pdf.
IEA. "Ocean Power." https://www.iea.org/reports/ocean-power.

IUCN. "Coral Reefs and Climate Change." (2017). https://www.iucn.org/sites/dev/files/coral_reefs_and_climate_change_issues_brief_final.pdf.

———. "Ocean Deoxygenation." https://www.iucn.org/resources/issues-briefs/ocean-deoxygenation.

IUCN, University of Cambridge, and The Nature Conservancy. "Global Assessments of Mangrove Losses and Degradation." (2018). https://www.iucn.org/sites/dev/files/content/documents/mangroveloss-brief-4pp-19.10.low_.pdf.

Izaguirre, C., I. J. Losada, P. Camus, J. L. Vigh, and V. Stenek. "Climate Change Risk to Global Port Operations." *Nature Climate Change* 11, no. 1 (2021/01/01/ 2021): 14–20.

Kirezci, Ebru, Ian R. Young, Roshanka Ranasinghe, Sanne Muis, Robert J. Nicholls, Daniel Lincke, and Jochen Hinkel. "Projections of Global-Scale Extreme Sea Levels and Resulting Episodic Coastal Flooding over the 21st Century." *Scientific Reports* 10, no. 1 (2020).

Komoroske, Lisa M., and Rebecca L. Lewison. "Addressing Fisheries Bycatch in a Changing World." [In English]. *Frontiers in Marine Science* 2, no. 83 (2015/10/27/ 2015).

Lebreton, Laurent C. M., Joost van der Zwet, Jan-Willem Damsteeg, Boyan Slat, Anthony Andrady, and Julia Reisser. "River Plastic Emissions to the World's Oceans." *Nature Communications* 8, no. 1 (2017/06/07/ 2017): 15611.

Marsay, Chris M., Richard J. Sanders, Stephanie A. Henson, Katsiaryna Pabortsava, Eric P. Achterberg, and Richard S. Lampitt. "Attenuation of Sinking Particulate Organic Carbon Flux through the Mesopelagic Ocean." *Proceedings of the National Academy of Sciences* (2015): 201415311.

Martinez-Porchas, Marcel, and Luis R. Martinez-Cordova. "World Aquaculture: Environmental Impacts and Troubleshooting Alternatives." [In eng]. *TheScientificWorldJournal* 2012 (2012): 389623–23.

Ministry for Primary Industries. "Overview of Ecological Effects of Aquaculture." (2013). https://www.mpi.govt.nz/dmsdocument/4300/direct.

Ministry for the Environment. "Adapting to Sea-Level Rise." https://www.mfe.govt.nz/climate-change/climate-change-and-government/adapting-climate-change/adapting-sea-level-rise.

NIWA. "Climate Change and the Oceans." https://niwa.co.nz/education-and-training/schools/students/climate-change/oceans.

NOAA. "Coral Reef Ecosystems." https://www.noaa.gov/education/resource-collections/marine-life/coral-reef-ecosystems#:~:text=Benefits%20of%20c oral%20reef%20ecosystems,food%2C%20income%2C%20and%20protection.

———. "Fast Facts Coral Reefs." https://coast.noaa.gov/states/fast-facts/coral-reefs.html

———. "How Do Coral Reefs Benefit the Economy?" https://oceanservice.noaa.gov/facts/coral_economy.html.

———. "What Is Coral Bleaching?" https://oceanservice.noaa.gov/facts/coral_bleach.html.

———. "What Is Ocean Acidification?" https://www.pmel.noaa.gov/co2/story/What+is+Ocean+Acidification%3F.

OES-Environmental. "Oes-Environmental 2020 State of the Science Report: Environmental Effects of Marine Renewable Energy Development around the World." (2020). https://tethys.pnnl.gov/publications/state-of-the-science-2020.

Pörtner, H.-O., D.M. Karl, P.W. Boyd,W.W.L. Cheung, S.E. Lluch-Cota, Y. Nojiri, D.N. Schmidt, and P.O. Zavialov,. "Ocean Systems." *Climate Change 2014: Impacts, Adaptation, and Vulnerability. Part A: Global and Sectoral Aspects. Contribution of Working Group II to the Fifth Assessment Report of the Intergovernmental Panel on Climate Change* (2014). https://www.ipcc.ch/site/assets/uploads/2018/02/WGIIAR5-Chap6_FINAL.pdf.

Rabalais, Nancy N., R. Eugene Turner, Robert J. Díaz, and Dubravko Justić. "Global Change and Eutrophication of Coastal Waters." *ICES Journal of Marine Science* 66, no. 7 (2009): 1528–37.

Schmidtko, Sunke, Lothar Stramma, and Martin Visbeck. "Decline in Global Oceanic Oxygen Content During the Past Five Decades." *Nature* 542 (02/15/online 2017): 335.

ScienceDaily. "Ocean Sink for Human-Made Carbon Dioxide Measured."

Senko, J., E. R. White, S. S. Heppell, and L. R. Gerber. "Comparing Bycatch Mitigation Strategies for Vulnerable Marine Megafauna." *Animal Conservation* 17, no. 1 (2014/02/01/ 2014): 5–18.

Smithsonian Ocean. "Ocean Acidification." https://ocean.si.edu/ocean-life/invertebrates/ocean-acidification.

Song, Andrew M., Joeri Scholtens, Kate Barclay, Simon R. Bush, Michael Fabinyi, Dedi S. Adhuri, and Milton Haughton. "Collateral Damage? Small-Scale Fisheries in the Global Fight against Iuu Fishing." *Fish and Fisheries* 21, no. 4 (2020/07/01/ 2020): 831–43.

Sumaila, U. R., D. Zeller, L. Hood, M. L. D. Palomares, Y. Li, and D. Pauly. "Illicit Trade in Marine Fish Catch and Its Effects on Ecosystems and People Worldwide." *Science Advances* 6, no. 9 (2020): eaaz3801.

Thiel, Martin, Guillermo Luna-Jorquera, Rocío Álvarez-Varas, Camila Gallardo, Iván A. Hinojosa, Nicolás Luna, Diego Miranda-Urbina, *et al.* "Impacts of Marine Plastic Pollution from Continental Coasts to Subtropical Gyres—Fish, Seabirds, and Other Vertebrates in the Se Pacific." [In English]. *Frontiers in Marine Science* 5, no. 238 (2018/07/24/ 2018).

UN. "Factsheet: Biodiversity." (2017). https://sustainabledevelopment.un.org/content/documents/Ocean_Factsheet_Biodiversity.pdf.

UNCTAD. "The Covid-19 Pandemic and the Blue Economy: New Challenges and Prospects for Recovery and Resilience." (2020). https://unctad.org/en/PublicationsLibrary/ditctedinf2020d2_en.pdf.

UNEP. "Emissions Gap Report 2019." (2019). https://www.unenvironment.org/resources/emissions-gap-report-2019.

———. "Land-Based Pollution." https://www.unenvironment.org/explore-topics/oceans-seas/what-we-do/working-regional-seas/land-based-pollution.

UNESCO. "Facts and Figures on Marine Biodiversity." http://www.unesco.org/new/en/natural-sciences/ioc-oceans/focus-areas/rio-20-ocean/blueprint-for-the-future-we-want/marine-biodiversity/facts-and-figures-on-marine-biodiversity/.

Waycott, Michelle, Carlos M. Duarte, Tim J. B. Carruthers, Robert J. Orth, William C. Dennison, Suzanne Olyarnik, Ainsley Calladine, *et al.* "Accelerating Loss of Seagrasses across the Globe Threatens Coastal Ecosystems." *Proceedings of the National Academy of Sciences* (2009): pnas.0905620106.

Woodhead, Anna J., Christina C. Hicks, Albert V. Norström, Gareth J. Williams, and Nicholas A. J. Graham. "Coral Reef Ecosystem Services in the Anthropocene." *Functional Ecology* 33, no. 6 (2019/06/01/ 2019): 1023–34.

World Bank. "The Potential of the Blue Economy: Increasing Long-Term Benefits of the Sustainable Use of Marine Resources for Small Island Developing States and Coastal Least Developed Countries." (2017). https://openknowledge.worldbank.org/bitstream/handle/10986/26843/115545.pdf?sequence=1&isAllowed=y.

Worm, Boris, Edward B. Barbier, Nicola Beaumont, J. Emmett Duffy, Carl Folke, Benjamin S. Halpern, Jeremy B. C. Jackson, *et al.* "Impacts of Biodiversity Loss on Ocean Ecosystem Services." *Science* 314, no. 5800 (2006): 787.

WWF. "Living Blue Planet Report." (2015). http://ocean.panda.org.s3.amazonaws.com/media/Living_Blue_Planet_Report_2015_08_31.pdf.

———. "Living Planet Report 2020." (2020). https://livingplanet.panda.org/en-us/.

3

The Blue Economy

Introduction

In the twenty-first century, coastal countries have begun to consider the ocean a source of jobs, innovation, and competitive advantage. Nonetheless, the ocean's value is reduced by numerous environmental pressures. With the ocean being the driver of economic growth, there are calls from multilateral agencies, scientists, and the public to ensure that maritime industries and the use of ocean space, resources, and ecosystems are ecologically sustainable and that economic activities are in balance with the long-term carrying capacity of the ocean ecosystems.[1,2] This chapter will first introduce the concept of the blue economy and blue growth. Following this, the chapter will discuss the blue economy in the context of sustainable development. Finally, the chapter will review actions to guide the development of the blue economy.

R. C. Brears, *Developing the Blue Economy*,
https://doi.org/10.1007/978-3-030-84216-1_3

Blue Economy

The ocean economy is mainly driven by growth in aquaculture, offshore wind, fish processing, and shipbuilding and repair, with the value of critical ocean assets estimated to be around $24 trillion and the value derived from services estimated to be $2.5 trillion per annum, equivalent to 3–5% of global gross domestic product.[3,4,5] Nonetheless, the ocean's value is reduced by environmental pressures from overfishing, climate change, pollution, loss of habitats and biodiversity, and coastal development. In response, there are calls for developing a blue economy that ensures economic development does not come at the detriment of the ocean and its ecosystems.

The blue economy can be defined "*as comprising the range of economic sectors and related policies that together determine whether the use of oceanic resources is sustainable. An important challenge of the blue economy is thus to understand and better manage the many aspects of oceanic sustainability, ranging from sustainable fisheries to ecosystem health to pollution*". In this context, the blue economy seeks to promote economic growth, social inclusion, and the preservation or improvement of livelihoods while at the same time ensuring the sustainability of the oceans and coastal areas.[6] Furthermore, the conservation of ocean ecosystems and biological resources can generate multiple economic, environmental, and social benefits to individual sectors and society by delivering broader ecosystem services and increased human well-being. For example, coral ecosystems are estimated to have an annual value of $172 billion based on the ecosystem services they provide, including food and raw materials, moderation of extreme ocean events, water purification, recreation, tourism, and maintenance of biodiversity.[7,8]

The blue economy comprises established and emerging sectors and practices, including sustainable fishing and aquaculture, marine biotechnology, marine renewable energy, coastal water resources management, blue carbon initiatives, and ecosystem-based adaptation.[9,10] While the activities in each sector and practice may differ across countries, they focus on:

- Providing social and economic benefits for current and future generations
- Restoring, protecting, and maintaining the diversity, productivity, resilience, core functions, and intrinsic value of marine ecosystems
- Promoting clean technologies, renewable energy, and circular economy principles of decoupling economic growth from resource consumption, reducing waste, and recycling materials[11]

Blue Growth

Blue growth is a strategy for sustainable economic growth and job creation to reduce poverty while addressing resource scarcity and climate extremes. The World Bank defines blue growth as "*growth that is efficient in its use of natural resources, clean in that it minimises pollution and environmental impacts, and resilient in that it accounts for natural hazards and the role of environmental management and natural capital in preventing physical disasters*". The European Commission has defined blue growth as "*smart, sustainable and inclusive economic and employment growth from the oceans, seas and coasts*".

While it is not clear as to what a sustainable blue economy should look like and the conditions that are the most favourable for its development, blue growth aims to promote the development of the maritime economic functions in a sustainable manner, where the maritime economy consists of all the sectoral and cross-sectoral economic activities that relate to the oceans, seas, and coasts and includes direct and indirect supporting activities necessary for the functioning of the maritime economic sectors while maritime employment is all the employment resulting from these activities related to the oceans, seas, and coasts.[12,13]

Blue growth focuses on technology innovations that contribute to economic development while ensuring the sustainable management of natural, marine resources. In addition to blue growth ensuring the economic optimisation or more effective use of resources, it also involves social innovation, which is the changes of attitudes, behaviour, and perception of groups of people towards the ocean and its resources.[14] Finally, blue growth recognises that diverse ocean uses by humans (such as fisheries and shipping) and marine ecosystem services (for example, food provisioning, coastal protection, and carbon storage) are interconnected. Therefore, synergies can be gained, and trade-offs minimised by managing human uses and ecosystem services jointly, rather than in isolation. A failure to manage uses and services jointly can lead to sub-optimal outcomes at the system level, even if all the components are managed for the same objective.[15]

Technologies and Blue Growth

It is becoming increasingly common to distinguish between three categories of technologies for environmental management, which can be applied in the blue economy: hardware, software, and orgware. Hardware refers to "hard" technologies, such as capital goods and equipment. Software refers to the capacity and processes involved in using the technology and spans knowledge and skills, including awareness-raising, education, and training. Orgware refers to the institutional set-up and coordination mechanisms required to support hardware and software implementation and includes the organisational, ownership, and institutional arrangements necessary for successful implementation and sustainability of blue economy solutions.[16,17]

Blue Economy and Sustainable Development

The blue economy concept does not replace that of sustainable development; instead, it can be understood as a way to achieve sustainable development, which is "*development that meets the needs of the present without compromising the ability of future generations to meet their own needs*".[18,19] Consisting of three pillars, sustainable development seeks to achieve, in a balanced manner, economic, social, and environmental sustainability.

Economic Sustainability

Weak sustainability assumes that natural capital and manufactured capital are essentially substitutable and that there are no essential differences between the kinds of well-being they create: the only thing that matters is the total value of the aggregate stock of capital, which should be at least maintained or ideally increased for the sake of future generations. Also, in the weak form of sustainability, the economy will continually generate technical solutions to environmental problems caused by increased production levels of goods and services. In contrast, the strong sustainability view sees natural capital as a complex system that consists of evolving biotic and abiotic elements that interact in ways that determine the ecosystem's capacity to provide ecosystem services. In this context, strong sustainability proponents state that natural capital is non-substitutable as natural capital cannot be reproduced. Its destruction is irreversible, in contrast to manufactured capital that can be reproduced and restored. Due to the lack of knowledge about ecosystems' functioning, society cannot be sure about the effects on human well-being from destroying natural capital. As such, acknowledging irreversibility and uncertainties should lead to implementing the precautionary principle regarding the use of natural capital in economic growth.[20]

Social Sustainability

An unjust society is unlikely to be sustainable in environmental or economic terms. Social tensions are likely to undermine citizens' recognition of both their environmental rights and duties relating to environmental degradation. Therefore, a better understanding of sustainable development's concept of social sustainability is critical for reconciling the competing demands of the society-environmental-economic tripartite. There are five interconnected equity principles of social sustainability listed in Table 3.1.[21,22,23]

Table 3.1 Equity Principles of Social Sustainability

Principle	Description
Intergenerational equity	This is the equity between generations where future generation's standards of living should not be disadvantaged by the activities of the current generation's standard of living
Intragenerational equity	This is the equity among the current generation which can be achieved through widespread political participation by citizens
Geographical equity (trans-frontier responsibility)	Local policies should be geared towards resolving local and global environmental problems as political/administrative boundaries are frequently used to shield polluters from prosecution in other jurisdictions
Procedural equity	Regulatory systems should be devised to ensure transparency as people must have the right to access environmental information on activities that have both local and global impacts
Interspecies equity	This notion places the survival of other species on an equal basis to the survival of humans. This is to reflect the critical importance of preserving ecosystems and maintaining biodiversity for human survival. Specifically, humans must ensure ecosystems are not degraded beyond their regenerative capacity

Environmental Sustainability

Environmental sustainability is the ability to maintain and enhance the physical environment's qualities, for instance, maintaining the living conditions for people and other species. Specifically, environmental sustainability aims to protect natural ecosystems' integrity and the various ecosystem services necessary for human survival. There are four types of ecosystem services that nature provides, with examples of benefits in the context of the blue economy summarised in Table 3.2:

Table 3.2 The Blue Economy's Main Ecosystem Services

Ecosystem Service	Capacity	Example of Benefits
Provisioning services	Food provision	Biomass for human consumption
	Water storage and provision	Water for human consumption and other uses
	Biotic materials and biofuels	Biomass or biotic elements for non-food purposes
Regulating services	Water purification	Bio and physicochemical processes for waste and pollutant removal
	Air quality regulation	Habitats help to intercept or absorb airborne pollutants
	Coastal protection	Erosion prevention, protection against floods, storms, etc
	Climate regulation	Uptake, storage, and sequestration of carbon dioxide
	Weather regulation	Influence on local weather conditions as thermoregulation and humidity
	Ocean nourishment	Soil formation and quality regulation
	Life cycle maintenance	Biological and physical support to healthy species and their reproduction
	Biological regulation	Biological control of pests that may affect commercial activities and human health
Supporting services	Photosynthesis	Phytoplankton are photosynthetic and form the basis of the marine food web, and are responsible for carbon dioxide fixation

(continued)

Table 3.2 (continued)

Ecosystem Service	Capacity	Example of Benefits
	Nutrient cycling	The movement and exchange of organic and inorganic matter back into the production of matter, with nutrients dissolved in seawater essential for the survival of marine life
	Marine sediments	The sediments provide habitat for a multitude of marine life, particularly marine microorganisms
	Water cycle	Oceans provide evaporated water to the water cycle and allow water to move all around the globe as ocean currents
Cultural services	Symbolic and aesthetic values	Exaltation of senses and emotions by seascapes, habitats, and species
	Recreation and tourism	Opportunities for relaxation and amusement
	Cognitive effects	Trigger of mental process like knowing, developing, perceiving

- Provisioning services are products obtained from ecosystems
- Regulating services are benefits obtained from the regulation of ecosystem processes
- Supporting services that are necessary for the continuation of the three above-mentioned ecosystem services types
- Cultural services, which are non-material benefits obtained from ecosystems such as religious, spiritual, recreational, and educational[24,25]

Conflict Prevention and Mitigation

The maritime sectors currently utilising marine space are diverse, ranging from global international sectors, for example, maritime transport, to more regional sectors such as aquaculture and maritime tourism. These sectors usually have different levels of development in national and regional economies. Each maritime sector uses space in different ways. Hard sectors require fixed infrastructure in the sea, for example, aquaculture installations or offshore wind farms. These structures are typically in place for a long time, are expensive to install, and cannot be easily moved once established. Soft sectors are those that do not use fixed infrastructure, such as tourism or fishing. They tend to be more fleeting and less fixed in terms of space as they need to respond to changing environmental conditions. Both hard and soft sectors usually look for ideal sites that offer the best conditions for their activities, with preferred locations usually chosen based on various environmental, economic, and technological factors, including water depth or distance to shore. As such, there is potential for conflict between sectors.[26]

Generally, conflicts arise because sectors' activities come into conflict with each other, with the conflicts themselves typically described as sectoral. Cross-sectoral spatial conflicts arise from direct competition over limited space or one sector negatively impacting the other, which may or may not be in the same location. Conflicts can occur at different spatial scales, sometimes even concurrently. Conflicts involving local communities are generally local or regional as they tend to be tied to particular locations or local/regional economies. Other conflicts occur at the national or international level, usually involving international activities. Conflicts related to protected spaces and habitats are also often transnational. Meanwhile, nearshore conflicts are different from offshore conflicts: Nearshore conflicts tend to be more immediate and tangible for stakeholders and local communities, while offshore conflicts are generally more specialised, only involve the respective sectors, and are out of sight from local communities. Table 3.3 lists a few potential drivers that can escalate conflicts.[27]

There are two main options for addressing conflicts in the development of the blue economy:

Table 3.3 Potential Drivers of Spatial Conflicts

Driver	Description
Political priorities	Political priorities may change, with one sector having a higher priority placed on it compared to another
Stakeholder perceptions and lack of understanding	Lack of understanding of how a sector works can impede constructive discussions. It can also prevent solutions from being found if mutual needs and the reasons for those needs are not made clear
Lack of transparency of a decision-making process	Providing stakeholders with incomplete information may lead to questions or suspicions and potentially even derail or delay the process due to a breakdown in trust
Spatial constraints	Spatial constraints can restrict spatial management options such as relocation
Media exposure	Media exposure can bring stakeholders into dialogue, but it can also escalate a conflict that might otherwise find a solution. The use of the media for political gain can also escalate conflicts
Lack of knowledge or contested knowledge	Lack of knowledge or contested knowledge of activities' impacts can escalate conflicts, especially where environmental impacts are concerned. Uncertainties can be challenging to deal with in decision-making processes, and knowledge can be contested
Lack of resources, time, and clear responsibilities	A lack of resources, time, and clear responsibilities can significantly escalate a conflict
Lack of acceptance of the proposed solution	Some conflicts can escalate because a small number of stakeholders are unwilling to accept the solution

- *Conflict prevention*: Conflict prevention is action that seeks to avert spatial competition, usually by ensuring that incompatible activities do not occur in the same space or negatively affect each other. Conflict prevention takes place before conflict occurs
- *Conflict mitigation*: Conflict mitigation is action that seeks to reduce the impacts of spatial competition. It is relevant for unavoidable conflicts, for instance, because spatial options are already congested

Stakeholders (sector stakeholders, public stakeholders, and local communities) play many roles in spatial conflict resolution. Stakeholders can be crucial for bringing conflicts to the table for resolution, particularly in smaller or less prominent sectors. They are also essential as conflict managers and designers of solutions. Stakeholder acceptance is essential for implementing solutions, especially mitigation, where many solutions will depend on the stakeholders' voluntary commitments. An important aspect to note is that no two sectors are the same, with some being fragmented and underfunded while others are organised and have substantial financial resources.[28]

Principles of the Blue Economy and Blue Growth

Overall, a set of principles are detailed in Table 3.4 that define the blue economy and blue growth.[29]

Actions to Guide the Development of the Blue Economy

There are a set of actions that can foster the development of the blue economy that maximises synergies and reduces trade-offs, including the following:

Table 3.4 Principles of the Blue Economy and Blue Growth

Principle	Description
Provides social and economic benefits for current and future generations	The blue economy contributes to food security, poverty alleviation, incomes and livelihoods, employment, health, equity, and political stability
Maintains marine ecosystems	The blue economy restores, protects, and maintains the diversity, productivity, resilience, core functions, and intrinsic value of marine ecosystems—the natural capital upon which its prosperity depends
Promotes clean technologies and renewable energy	The blue economy is based on clean technologies, renewable energy, and circular material flows to secure economic and social stability over time while keeping within the limits of one planet
Inclusive	The blue economy is centred around active and effective stakeholder engagement and participation
Informed	The blue economy is guided by scientifically sound information to prevent harmful effects that undermine long-term sustainability. When information is lacking, actors take a precautionary approach. When new knowledge of risks and sustainable opportunities is gained, actors adapt their decisions and activities
Accountable and transparent	In the blue economy, actors take responsibility for the impacts of their activities by taking appropriate actions as well as being transparent about their impacts so stakeholders are well-informed and can exert their influence
Holistic, cross-sectoral, and long term	In the blue economy, decisions consider economic, social, and environmental costs and benefits to society as well as their impacts on other activities now and in the future

(continued)

Table 3.4 (continued)

Principle	Description
Innovative and proactive	The blue economy is based on innovative ways to meet the needs of present and future generations without undermining the capacity of nature to support human economic activities and well-being

- *Define objectives, quantify trade-offs, and strive for efficiency*: Managing complex systems involves making trade-offs between conflicting stakeholder objectives. It is vital to have a clear understanding of each stakeholder's objectives and how they interact. This enables quantification and communication of trade-offs between objectives so that managers and stakeholders understand the full range of management options and avoid false trade-offs and conflicts when synergies can be created, leading to win–win situations. Efficiency is a good first target when managing multiple objectives that are difficult to compare objectively. With regard to trade-offs, efficiency means an outcome has been achieved from which it is not possible to improve one objective without regressing in at least one other
- *Maximise existing data*: Data limitation is often regarded as an obstacle to the science-based management of complex systems. It is challenging to design improvements to systems if the current state is poorly understood, and it is challenging to assess trade-offs if the data required is unavailable. The more complex the system being managed, the greater the information required for management in terms of human uses and ecological interactions. Given the cost and capacity required to collect data, science-based management should focus on optimising the use of existing data and should evaluate the need for new data within a value-of-information framework. Advances in data science and data-limited assessment demonstrate that additional information can be extracted out of existing data:

  – *If two systems are similar, information from one has value in the other*: While the world is complex, similarities between systems often allow data from one system to carry information that has value in others. Statistical correlations can be conducted to explore relationships, while natural experiments can extract information out of cross-system data

- *Measurement of a system component carries information about all components influencing it*: If a variable is influenced by other variables, then measurements of the state and dynamics of the first variable also carry information about the states and dynamics of other related variables
- *If multiple states or processes can be measured relative to one another, only one must be measured in absolute*: In data-poor systems, relatives are usually easier to measure than their corresponding absolutes
- *Utilise stakeholder's knowledge and citizen science*: Effective stakeholder engagement is essential for managing blue growth, particularly in data-poor and governance-poor contexts. Stakeholder's ecological knowledge, as well as citizen science, can provide a valuable means of accessing new data. By engaging with the public, researchers and managers can collect vastly more data, and the data produced by citizen scientists can be just as reliable
- *Measure impact for iterative learning*: Effectively measuring an intervention's impact requires careful planning before the intervention is undertaken. Measuring impacts provides three benefits. First, it allows scientists and managers to assess the efficacy of an intervention. Second, it may facilitate addressing broader scientific questions that can help guide future interventions in other locations. Third, the process of iterative learning and revising an intervention operationalises adaptive management

• *Design institutions, not behaviours*: Blue growth should focus on designing institutions instead of focusing on managers directly designing stakeholder behaviour. For example, institutions are social structures, such as regulations, norms, markets, under which stakeholders choose their behaviours. As such, managers do not directly control behaviours but instead influence behaviours by changing institutions. A key design feature of institutions is that they correct market failure and/or promote resilience[30]

Notes

1. G20 Insights, "Sustainable Ocean Economy, Innovation and Growth: A G20 Initiative for the 7th Largest Economy in the World." (2017), https://www.g20-insights.org/policy_briefs/sustainable-ocean-economy-innovation-growth-g20-initiative-7th-largest-economy-world/.
2. OECD, "Oecd Work in Support of a Sustainable Ocean," (2020), https://www.oecd.org/ocean/OECD-work-in-support-of-a-sustainable-ocean.pdf.
3. Luca Mulazzani and Giulio Malorgio, "Blue Growth and Ecosystem Services," *Marine Policy* 85 (2017).
4. OECD, "Oecd Work in Support of a Sustainable Ocean."
5. G20 Insights, "Sustainable Ocean Economy, Innovation and Growth: A G20 Initiative for the 7th Largest Economy in the World."
6. World Bank, "The Potential of the Blue Economy: Increasing Long-Term Benefits of the Sustainable Use of Marine Resources for Small Island Developing States and Coastal Least Developed Countries," (2017), https://openknowledge.worldbank.org/bitstream/handle/10986/26843/115545.pdf?sequence=1&isAllowed=y.
7. G20 Insights, "Sustainable Ocean Economy, Innovation and Growth: A G20 Initiative for the 7th Largest Economy in the World."
8. OECD, "Oecd Work in Support of a Sustainable Ocean."
9. Andrew D. L. Steven, Mathew A. Vanderklift, and Narnia Bohler-Muller, "A New Narrative for the Blue Economy and Blue Carbon," *Journal of the Indian Ocean Region* 15, no. 2 (2019).
10. R.C. Brears, *Nature-Based Solutions to 21st Century Challenges* (Oxfordshire, UK: Routledge, 2020).
11. World Bank, "The Potential of the Blue Economy: Increasing Long-Term Benefits of the Sustainable Use of Marine Resources for Small Island Developing States and Coastal Least Developed Countries."
12. Ibid.
13. Ecorys, "Blue Growth Scenarios and Drivers for Sustainable Growth from the Oceans, Seas and Coasts," (2012), https://ec.europa.eu/

maritimeaffairs/sites/maritimeaffairs/files/docs/publications/blue_growth_third_interim_report_en.pdf.
14. Katrine Soma et al., "Social Innovation – a Future Pathway for Blue Growth?," *Marine Policy* 87 (2018).
15. Matthew G. Burgess et al., "Five Rules for Pragmatic Blue Growth," ibid.
16. UNEP-DTU UNEP-DHI Partnership, CTCN, "Climate Change Adaptation Technologies for Water: A Practitioner's Guide to Adaptation Technologies for Increased Water Sector Resilience," (2017), https://www.ctc-n.org/resources/climate-change-adaptation-technologies-water-practitioner-s-guide-adaptation-technologies.
17. UNFCCC, "Draft Joint Policy Brief on Technologies on Technologies for Averting, Minimizing and Addressing Loss and Damage in Coastal Zones," (2020), https://unfccc.int/ttclear/misc_/StaticFiles/gnwoerk_static/tn_meetings/e6d08ac68eab42ab863d9dfe807964bd/cfb1b45c79b5400897a10a142a6bb10a.pdf.
18. Lu Wenhai et al., "Successful Blue Economy Examples with an Emphasis on International Perspectives," *Frontiers in Marine Science* 6, no. 261 (2019).
19. Brundtland Commission, "Our Common Future: Report of the World Commission on Environment and Development," *UN Documents Gatheringa Body of Global Agreements* (1987).
20. J. Pelenc, J. Ballet, and T. Dedeurwaerdere, "Weak Sustainability Versus Strong Sustainability," *Brief for GSDR United Nations* (2015), https://sustainabledevelopment.un.org/content/documents/6569122-Pelenc-Weak%20Sustainability%20versus%20Strong%20Sustainability.pdf.
21. Graham Haughton, "Environmental Justice and the Sustainable City," *Journal of Planning Education and Research* 18, no. 3 (1999).
22. Suzanne Vallance, Harvey C. Perkins, and Jennifer E. Dixon, "What Is Social Sustainability? A Clarification of Concepts," *Geoforum* 42, no. 3 (2011).
23. R.C. Brears, *The Green Economy and the Water-Energy-Food Nexus* (London: Palgrave Macmillan UK, 2017).
24. Ibid.

25. European Commission, "The Eu Blue Economy Report 2020," (2020), https://ec.europa.eu/maritimeaffairs/sites/maritimeaffairs/files/2020_06_blueeconomy-2020-ld_final.pdf.
26. "Addressing Conflicting Spatial Demands in Msp: Considerations for Msp Planners," (2019), https://www.msp-platform.eu/sites/default/files/20190604_conflicts_study_published_0.pdf.
27. Ibid.
28. Ibid.
29. WWF, "Principles for a Sustainable Blue Economy," (2016), https://unctad.org/meetings/en/Contribution/ditc-ted-oceans-wwf-blue-economy-2016.pdf.
30. Burgess et al., "Five Rules for Pragmatic Blue Growth."

References

Brears, R. C. *The Green Economy and the Water-Energy-Food Nexus.* London: Palgrave Macmillan UK, 2017.

———. *Nature-Based Solutions to 21st Century Challenges.* Oxfordshire, UK: Routledge, 2020.

Brundtland Commission. "Our Common Future: Report of the World Commission on Environment and Development." *UN Documents Gatheringa Body of Global Agreements* (1987).

Burgess, Matthew G., Michaela Clemence, Grant R. McDermott, Christopher Costello, and Steven D. Gaines. "Five Rules for Pragmatic Blue Growth." *Marine Policy* 87 (2018/01/01/ 2018): 331–39.

Ecorys. "Blue Growth Scenarios and Drivers for Sustainable Growth from the Oceans, Seas and Coasts." (2012). https://ec.europa.eu/maritimeaffairs/sites/maritimeaffairs/files/docs/publications/blue_growth_third_interim_report_en.pdf.

European Commission. "Addressing Conflicting Spatial Demands in Msp: Considerations for Msp Planners." (2019). https://www.msp-platform.eu/sites/default/files/20190604_conflicts_study_published_0.pdf.

———. "The Eu Blue Economy Report 2020." (2020). https://ec.europa.eu/maritimeaffairs/sites/maritimeaffairs/files/2020_06_blueeconomy-2020-ld_final.pdf.

G20 Insights. "Sustainable Ocean Economy, Innovation and Growth: A G20 Initiative for the 7th Largest Economy in the World." (2017). https://www.g20-insights.org/policy_briefs/sustainable-ocean-economy-innovation-growth-g20-initiative-7th-largest-economy-world/.

Haughton, Graham. "Environmental Justice and the Sustainable City." *Journal of Planning Education and Research* 18, no. 3 (1999): 233–43.

Mulazzani, Luca, and Giulio Malorgio. "Blue Growth and Ecosystem Services." *Marine Policy* 85 (2017/11/01/ 2017): 17–24.

OECD. "Oecd Work in Support of a Sustainable Ocean." (2020). https://www.oecd.org/ocean/OECD-work-in-support-of-a-sustainable-ocean.pdf.

Pelenc, J., J. Ballet, and T. Dedeurwaerdere. "Weak Sustainability Versus Strong Sustainability." *Brief for GSDR United Nations* (2015). https://sustainabledevelopment.un.org/content/documents/6569122-Pelenc-Weak%20Sustainability%20versus%20Strong%20Sustainability.pdf.

Soma, Katrine, Sander W. K. van den Burg, Ellen W. J. Hoefnagel, Marian Stuiver, and C. Martijn van der Heide. "Social Innovation – A Future Pathway for Blue Growth?" *Marine Policy* 87 (2018/01/01/ 2018): 363–70.

Steven, Andrew D. L., Mathew A. Vanderklift, and Narnia Bohler-Muller. "A New Narrative for the Blue Economy and Blue Carbon." *Journal of the Indian Ocean Region* 15, no. 2 (2019/05/04/ 2019): 123–28.

UNEP-DHI Partnership, UNEP-DTU, CTCN,. "Climate Change Adaptation Technologies for Water: A Practitioner's Guide to Adaptation Technologies for Increased Water Sector Resilience." (2017). https://www.ctc-n.org/resources/climate-change-adaptation-technologies-water-practitioner-s-guide-adaptation-technologies.

UNFCCC. "Draft Joint Policy Brief on Technologies on Technologies for Averting, Minimising and Addressing Loss and Damage in Coastal Zones." (2020). https://unfccc.int/ttclear/misc_/StaticFiles/gnwoerk_static/tn_meetings/e6d08ac68eab42ab863d9dfe807964bd/cfb1b45c79b5400897a10a142a6bb10a.pdf.

Vallance, Suzanne, Harvey C. Perkins, and Jennifer E. Dixon. "What Is Social Sustainability? A Clarification of Concepts." *Geoforum* 42, no. 3 (2011/06/01/ 2011): 342–48.

Wenhai, Lu, Caroline Cusack, Maria Baker, Wang Tao, Chen Mingbao, Kelli Paige, Zhang Xiaofan, *et al.* "Successful Blue Economy Examples with an

Emphasis on International Perspectives." [In English]. *Frontiers in Marine Science* 6, no. 261 (201906/07/ 2019).

World Bank. "The Potential of the Blue Economy: Increasing Long-Term Benefits of the Sustainable Use of Marine Resources for Small Island Developing States and Coastal Least Developed Countries." (2017). https://openknowledge.worldbank.org/bitstream/handle/10986/26843/115545.pdf?sequence=1&isAllowed=y.

WWF. "Principles for a Sustainable Blue Economy." (2016). https://unctad.org/meetings/en/Contribution/ditc-ted-oceans-wwf-blue-economy-2016.pdf.

4

Sustainable Fisheries

Introduction

Fish play a vital role in reducing hunger and improving nutrition, with fish being a source of proteins and nutrients and long-chain fatty acids, iodine, vitamins, minerals, and calcium. Fish provide around 3.3 billion people with almost 20% of their average per capita intake of animal protein. As well as providing crucial health benefits, fisheries generate employment opportunities and support livelihoods for people.[1] This chapter will first introduce the concept of sustainable fisheries before discussing the various economic and technology tools available to rebuild global fish stocks. The chapter will then discuss the ecosystem approach to fisheries and the concept of ecosystem-based fisheries management (EBFM).

Sustainable Fisheries

Traditional fisheries management has focused on single species sustainability for commercially valuable species. The maximum sustainable yield

R. C. Brears, *Developing the Blue Economy*,
https://doi.org/10.1007/978-3-030-84216-1_4

(MSY) is commonly used as a measure of sustainable fisheries management as this is the level at which the largest catch can be maintained indefinitely. When catch from a fishery is at or greater than the MSY and no sustainable management measures are implemented, the fishery is likely to head for depletion over time. An important aspect to note is that management targets, such as MSY are set for individual fisheries or fish stocks. While single-species management can be successful, a growing population and increased demand for seafood result in more than half of fisheries being exploited at their maximum level. Many do not have management measures to prevent over-exploitation. In addition, climate change and pollution are exacerbating the damage to fragile marine ecosystems. Therefore, additional or modified management objectives need to be adopted to ensure the sustainability of the broader marine system, such as targets and limits expressed in terms of yield or catch, the biomass of the fish stock remaining in the water, or measures of fish mortality. If sustainable fisheries practices are implemented, it could generate an additional $83 billion per annum while at the same time:

- *Sustaining livelihoods*: Sustainable fisheries provide safe, secure, and long-term employment, with people directly employed as fishers or indirectly employed in activities such as processing, marketing, and distribution. Furthermore, fishing is an activity that is strongly associated with the cultural identity and heritage of coastal communities
- *Ensuring food security*: The contribution of fish and fish products to global nutrition has ensured millions have avoided undernourishment, with fish catches, directly and indirectly, contributing to food security: directly, through the supply of the food commodity itself, and indirectly, when income obtained from those involved in the fishing industry is used to purchase food for families
- *Enhancing ecosystem resilience*: Rebuilding degraded fisheries helps increase their resilience and the resilience of the broader ecosystem to external shocks and stressors, including climate change and pollution. Sustainable fish stocks are crucial for ensuring healthy ecosystems, and healthy ecosystems are essential to the continued productivity of fisheries[2,3,4,5,6,7]

Tools to Rebuild Global Fish Stocks

There are various economic and technology tools available to rebuild global fish stocks, including the following.

Economic Tools

A key driver of the global fisheries decline is the unrestricted adoption of more powerful technology and a rapid increase in global fishing capacity leading to the over-exploitation of this renewable resource. There are a variety of economic tools that can restrict effort (the level of physical effort applied to a fishery to extract a certain amount of fish) and establish access rights to fisheries to ensure the long-term sustainability of fisheries, including the following:

- *Total Allowable Catch (TAC)*: Fisheries management aims to achieve the optimal and sustainable utilisation of the fishery resources to benefit humans while safeguarding the ecosystem. TAC establishes a level above which it becomes illegal to fish. If this level is set at the level of MSY or below and is well-enforced, it will immediately reduce effort and catch to a more sustainable level
- *Individual Quotas (IQ)*: IQs are individually allocated fishing rights programmes that allow secure, exclusive shares of the TAC to individual fishers or individual entities. IQs are associated with vessel licences or vessel ownership and are generally allocated based on historical track records. IQs may or may not be transferable independently of the vessel or its licence. Holders of IQs are held accountable for their share of the catch, which provides fishers more flexibility in running their business, such as the opportunity to fish whenever they choose, for example, in good weather and when fish prices are high. IQs are considered successful as they align the fisher's economic incentives with the health of fish stocks and provide more stability and predictability within a fishing year and over time
- *Individual Transferable Quotas (ITQ)*: In the ITQ system, fishers are allocated a catch quota, which they can trade with other participants

in the fishery. Expressly, ITQs explicitly limit the fish that a fleet can harvest from a fishery and assign tradable shares of the TAC to the participants in the fishery

- *Community Quotas (CQ)*: CQ's are allocated to a fisher group (a fisheries association, producer organisation, or port). It is up to the collective unit how the quota is allocated to and used by its members and to ensure compliance with the catch tonnage allocated. Nonetheless, this form of allocation is generally overseen by government quota managers
- *Territorial Use Rights in Fisheries (TURF)*: TURF designates access to a portion of a seabed or sea area rather than to the catch itself. Harvest rights in TURFs can range from privileges to fish in areas leased from the government to complete ownership over the delineated TURF area. The use of TURFs to control access to resources benefits the welfare of communities by increasing fishery sustainability as it eliminates the race to fish. TURF is most successful for sedentary species such as oysters and clams. TURFs can be paired with adjacent no-take reserves to increase fish biomass, diversity, and abundance, enhance resilience to climate change and natural disturbances, conserve biodiversity and critical habitats, and provide alternative income through tourism[8,9,10,11,12]

Gear Restrictions

Gear restrictions affect the type, characteristic, and operation of fishing gear. Some gears are prohibited outright to avoid increases in fishing capacity through increased efficiency, some unwanted impact on non-commercial sizes, species, or critical habitats, or an injection of new technology that could significantly modify exploitation rights' existing distribution. The regulation of gear characteristics such as minimum mesh size or dimensions of mouth opening of nets or traps is usually introduced to control fish mortality or some particular component of the resource, such as preventing the catching of juveniles of the target species or fishing of bycatch species. Gear restrictions can also be used to reduce the total catch by reducing the potential efficiency of the fisher.[13]

Reducing Bycatch

Bycatch can be defined as "*discarded catch of marine species and unobserved mortality due to a direct encounter with fishing vessels and gear*". Bycatch can negatively affect protected species by harming individuals, contributing to population declines, and impeding population recovery. Bycatch of fish can contribute to overfishing and impede efforts to rebuild fish stocks, or negatively affect economically and socially fishers and communities that rely on the economic benefits from a fishery or fish for food. Bycatch can alter the availability of predators and prey, impacting marine ecosystems and fish productivity, while bycatch of benthic species such as corals and sponges can damage habitats for fish and other species.[14] A variety of innovative technologies to reduce bycatch of juvenile and unwanted fish species are listed in Table 4.1.[15]

Table 4.1 Innovative Technologies to Reduce Bycatch

Technological Fix	How it Works
Turtle excluder devices	A large metal grid in the neck of a trawl net that physically excludes turtles from the base of trawl nets while allowing shrimp to be caught effectively
Tori (bird-scaring) lines	Keeps seabirds from baited hooks
Weighted lines	Sinks hooks faster out of reach of seabirds
Side-setting	Reduces the scavenging area by half
Line-setting devices	Places baited hooks immediately underwater
Circle hooks	Reduces frequency of deeply ingested hooks and limits gut perforation
Pingers	Acoustic devices that alert marine mammals to the presence of gillnets to prevent entanglement
Medina panels	Fine-mesh net aprons that reduce the probability of dolphin entanglement during net retrieval

Case: New Zealand's Fish Quota Management System

New Zealand's Quota Management System (QMS) guides the sustainable use of fisheries. Under the QMS, a yearly catch limit—the TAC—is set for every fish stock (a species of fish, shellfish, or seaweed from a particular area). New Zealand has 98 species, or species groups, divided into 642 individual fish stocks under the QMS. Fish stocks in the QMS are separated by Quota Management Areas (QMAs). For example, the snapper fishery is divided into six QMAs. These areas are based on administrative and biological factors for the species, for instance, how many fish there are in different regions, enabling individual sustainable catch limits to be set for each area. The TAC is set to allow the maximum sustainable catch from a fish stock while accounting for natural variation. An allowance is made for recreational and customary fishing and other fishing-related mortality. The remainder is the total allowable commercial catch (TACC), limiting the fish that commercial fishers can catch. Each year, quota owners get an Annual Catchment Entitlement (ACE), which is the right to catch a certain amount of fish stock during the fishing year. The amount of ACE that quota holders get varies, depending on the TACC set for that year. ACE can be bought or sold during the year, and commercial fishers must have enough ACE to cover the QMS fish they catch. A failure to do so will incur financial penalties.[16]

Marine Protected Areas

A marine protected area (MPA) is an area that is afforded greater protection than the surrounding waters for biodiversity conservation or fisheries management purposes. The IUCN recognises six different categories of MPAs, classified according to their objectives, ranging from fully protected areas (no-take zones where no extraction is permitted) to multiple-use areas (where a range of resource uses are allowed) (Table 4.2). MPAs can protect critical habitats or life stages of commercial species, such as spawning grounds or limiting bycatch by closing areas where bycatch rates are high. They can also be used for controlling fish mortality of sedentary species where fish population distribution and habitat preferences are known and may also be helpful in data-poor situations. MPAs can also allocate rights in specific locations to accrue benefits to certain users, such as traditional/subsistence fishing

Table 4.2 IUCN Categories of Protected Areas

Category	Description
1	Protected area managed mainly for science or wilderness protection (Strict Nature Reserve/Wilderness Area)
2	Protected area managed mainly for ecosystem protection and recreation (National Park)
3	Protected area managed mainly for conservation of specific natural features (Natural Monument or Feature)
4	Protected area managed mainly for conservation through management intervention (Habitat/Species Management Area)
5	Protected area managed mainly for landscape/seascape conservation and recreation (Protected Landscape/Seascape)
6	Protected area managed mainly for the sustainable use of natural ecosystems (Managed Resource Protection Area)

communities or resolve conflicts.[17] MPA effectiveness in ecological terms is typically measured by comparing values of ecological or biological measures. For example, comparing the sizes of organisms, density and biomass of fish assemblages, species richness, and live cover of benthic organisms in MPAs and adjacent unprotected areas and/or before and after an MPA is established.[18] MPA networks can be developed where two or more MPAs complement each other.

Marine Protected Area Networks

The IUCN defines an MPA network as "*a collection of individual MPAs or reserves operating cooperatively and synergistically, at various spatial scales, and with a range of protection levels designed to meet objectives that a single reserve cannot achieve*". A network of MPAs can offer protection to fish populations, for instance, protect highly mobile adults that congregate in specific locations for spawning and others that are more sedentary and restricted to a specific habitat yet interact with neighbouring fish populations and marine communities through mobile larvae. Networks of MPAs can support each other through connections between them, where such connections could be currents transporting fish eggs and larvae, therefore potentially adding to more sustainable fish populations.[19]

Case: Ivory Coast's First Marine Protected Area
The Ivory Coast has announced the creation of its first MPA. Covering 2,600 square kilometres off the coast of Grand-Béréby, the MPA is a "partially protected" area that includes an integrally protected zone closed to all activities and an eco-development zone that supports sustainable fisheries and ecotourism activities. The creation of the MPA relied on scientific data collected by a team that included the University of Exeter, supported by funding from the United Kingdom's Darwin Initiative and the Rainforest Trust. The research included participatory work with local communities to collect data on the waters' biodiversity and health, including underwater surveys of marine habitats and undocumented reefs.[20]

Locally Managed Marine Areas

A locally managed marine area (LMMA) is an area of nearshore waters and its associated coastal and marine resources that are largely or wholly managed at the local level by the coastal communities, land-owning groups, partner organisations, and/or collaborative government representatives who reside or are based in the immediate area. While the objectives for establishing LMMAs may vary between communities and partner organisations, a broad set of objectives that LMMAs can achieve are summarised in Table 4.3. LMMAs differ from MPAs in that LMMAs are characterised by local ownership, use, and/or control, and in some areas, follow the region's traditional tenure and management practices. In contrast, MPAs are typically designated by a top-down approach with limited local input. Community-based adaptive management is the process through which LMMAs can be achieved and sustained. Community-based implies that management is carried out by, or with a significant role played by the community, local stakeholders, relevant user groups, and locally and nationally relevant institutions and private interests.[21,22,23]

Table 4.3 Objectives of Locally Managed Marine Areas

Overarching Objective	Specific Objectives
Increase long-term sustainable fisheries yield for income and food	• Increase catch abundance/size for sustainable livelihoods • Increase catch abundance/size for food security
Increase efficiency of harvests and recovery for fish and invertebrate populations for short-term gain/contingency needs	• Ensure adequate fish and invertebrates to meeting fundraising targets • Provide adequate stock for village feasts • Ensure resource availability as a contingency for bad weather • Decrease variability in the food supply • Ensure resource availability for unplanned community event/needs
Maintain/restore biodiversity, habitats, and ecosystem function and improve resilience	• Conserve biodiversity/species • Protect/restore habitats • Maintain/restore ecosystem functions (e.g., productivity, herbivory, water filtration) • Maintain/restore ecosystem services (e.g., coastal protection) • Maintain/restore ecosystem resilience • Adapt to climate change
Maintain/restore biomass and breeding populations of targeted species	• Protect habitat for sensitive life history stages (breeding and nursery grounds) • Increase the stock of target species • Maintain/increase the size for greater reproductive capacity
Enhance economy and livelihoods	• Earn income from participation in management (e.g., paid ranger position) • Earn income from access fees, ecotourism activities, and alternative livelihoods

(continued)

Table 4.3 (continued)

Overarching Objective	Specific Objectives
Maintain or reinforce customs and traditions	• Preserve traditional practice • Demonstrate stewardship • Protect sacred sites • Secure or enhance respect for traditional leaders • Enforce custom as part of cultural practice
Assert access rights	• Limit or exclude outsiders from accessing resources • Assert tenure rights
Increase community organisation, cohesiveness, and empowerment	• Provide equitable access to resources • Strengthen community governance • Strengthen community participation • Strengthen engagement with the community • Obtain access to information to support management decisions • Attract non-profit support and their resources

Case: Fiji's Locally Managed Marine Area Network

Throughout the Pacific, traditional fisheries management practices include establishing a tabu area, a reef area where all fishing is prohibited. These tabu areas are traditionally put in place for 100 nights after a prominent chief or village leader's death. After the tabu is lifted, the abundant fish are harvested for a feast to celebrate the passing of the senior member of the community. With increasing pressure on fisheries in the mid-1990s, the community in Verata, Tailevu, on Fiji's main island of Viti Levu, worked together with government and non-government partners to reinvigorate the traditional practice of tabu to promote the preservation, protection, and sustainable use of marine resources. Over 400 villages around Fiji are now working with government, and non-government organisation partners under the Fiji Locally Managed Marine Area (FLMMA) network. These villages all have one or more tabu areas within their *iQoliqoli* (customary fishing areas) and manage this together with other restrictions on what can be caught, when, and how. FLMMA network efforts have resulted in 135 of Fiji's *iQoliqoli* being managed and having over 450 fishing reserves or tabu areas covering over 1,000 square kilometres.[24,25]

Fishery Monitoring

Fishery monitoring is essential for allowing fisheries to reach their full potential for producing food, revenue, and jobs while protecting ocean ecosystems. Monitoring systems generate data for resource managers to ensure compliance with fishery regulations aimed at achieving these goals. Monitoring also generates data required for scientific stock assessments, which can set sustainable catch limits. Overall, by monitoring fishing activity, regulators determine the types and quantity of fish being caught, monitor bycatch, gather accurate and useful information to ensure that fishers follow regulations, and fine or prosecute where illegal activity happens.[26,27] Fishery monitoring can be conducted in numerous ways, including the following.

At-Sea Fishery Observer Programmes

Information on how fishing vessels are operating at sea provides fish management authorities with the ability to assess the state of fish stocks and the ecosystems of which they are part. Observers can collect information on fishing vessels or at landing places, processing plants, or market places. Observer programmes are usually implemented to generate data for both fishery science and vessel compliance purposes, which then serve broader fisheries management objectives, specifically:

1. *Fisheries science*: This involves the estimation of total catch and effort, including bycatch, discards, biological sampling of catches, for

example, spawning condition, fish size, disease rates, shell condition, etc., and measurement of environmental parameters
2. *Vessel compliance*: Observers can register compliance with fisheries management laws, regulations, and plans, record catch composition, prohibited species, bycatch, size limits, discarding, area and gear restrictions, and validate vessel logbooks and the labelling of processed fish
3. *Effective fisheries management*: Both scientific and compliance types of data are needed for effective fisheries management. For example, they are vital determinations of opening and closing of a fishery; MPAs and species; condition of caught and released species; estimating pollution levels and so forth[28]

Catch Documentation Schemes

Catch documentation schemes (CDS) are market-related measures that have been developed to specifically combat illegal, unreported, and unregulated (IUU) fishing.

CDS track and trace fish throughout the supply chain. They record and certify information that identifies the origin of the fish caught and ensures it was harvested according to relevant conservation and management regulations. The objective of CDS is to combat IUU fishing by restricting the access of IUU fish and fishery products to markets. For CDS schemes to be effective, they should follow a set of standards and functions as detailed in Table 4.4.[29,30]

Case: The European Union's Web-based Catch Certification Scheme
The European Union's (EU) catch certification scheme, introduced in 2008, aims to prevent, deter, and eliminate IUU fishing. It covers all marine wild-caught fish (with some minor exemptions) trade by non-EU countries into the EU market. CATCH is an IT system that aims to digitalise the currently paper-based EU catch certification scheme, with version 1.0 launched in 2019. Fishery consignments need to be accompanied by a Catch Certificate validated by the flag state of the catching

vessel. More than 90 flag states have notified their competent authorities to validate certificates of export of fishery products into the EU territory. The objective behind CATCH is to develop a web-based application to support the management (issuance, control, and verification) of official documents and to automate the related regulatory procedures.[31]

Table 4.4 Standards and Functions of Catch Documentation Schemes

Standards and Functions	Description
Definition of CDS objective	A CDS should have a clear objective and minimise the burden on users
Definition of traceability standards	A CDS should define the standard of traceability required to meet its objective. The standard should define which segments of the supply chain are covered and to what level of detail
Specified monitoring, control, and surveillance controls	A CDS should be supported using monitoring, control, surveillance tools, and other available information, including logbooks, vessel monitoring systems (VMS), observer data, etc
e-CDS	CDSs should be designed as centralised electronic systems (e-CDSs). The e-CDS should serve as the point of generation, issuance, validation, and verification of all catch and trade certificates and serve as the repository of CDS data. Also, e-CDS should minimise the burden on users throughout the supply chain
Data exchange and standards	A CDS should adhere to global standards of information exchange and data management. They should be designed to ensure interoperability between states and be based on agreed international standards and formats

(continued)

Table 4.4 (continued)

Standards and Functions	Description
Data input and validation	A CDS should define roles and responsibilities for data input and validation. The private sector should initiate CDS certificates by supplying all relevant data, validated by the competent authorities
Access control and defined user roles	Access to an e-CDS should be defined and controlled through a hierarchical system of user logins and passwords that determine which parts, functions, and levels of the system users or users' groups can access
e-CDS functions and functionalities	An e-CDS should provide functionality that allows users to interact with it through a flexible and user-friendly interface. Such functions may include, among others, routines for recalling and rectifying certificates, printing out certificates physically, uploading scanned documentation along with data submission, or allowing private- and public-sector users to query and analyse those data accessible to them
Timing of certificate issuance	A CDS should clearly define at what points in the supply chain the initiation of certificates should be triggered. No catch should move to the next step in the supply chain without being covered by a certificate to minimise gaps and avenues of fraud in the system
Document numbering	A CDS should have system-generated, unique document numbering

Electronic Monitoring

Electronic monitoring (EM) programmes monitor catch and discards to generate high-quality monitoring data in fisheries. EM programmes include technology to collect data at sea (cameras and other sensors), processes for data analysis, and the creation of data products, such as reports, to ensure compliance with catch and discard limits and document that compliance.

At-Sea Technology

At-sea technology necessary for EM consists of a control centre and sensors that monitor different aspects of fishing operations. Cameras record vessel activities and may provide wide panoramic or close-up views of certain areas of the vessel. Panoramic views may provide a good overall view of vessel activity but may not provide detail such as individual fish or measure lengths. In addition to cameras, EM systems generally have sensors to activate image recording when certain vessel operations occur and record fishing operations. These sensors include:

- *GPS receiver*: Delivers data on time, vessel position, speed, heading, and position fix quality
- *Hydraulic pressure transducer*: They are generally mounted on the hydraulic system supply for deck winches and fishing equipment, with pressure readings indicating when the equipment is activated
- *Drum rotation sensor*: This is a photoelectric sensor that detects the motion of rotating drums that spool fishing gear by sensing the reflective tape mounted on the drums
- *Radio Frequency Identification (RFID)*: This technology uses electromagnetic fields to read information stored on tags automatically. More recent UHF-based RFID provides a very cost-effective way to identify tags on gear at distances of more than five metres[32]

Drones

There is a growing use of fully or partly unmanned vehicles, or drones, for sustainable fisheries management. There are three types of drones: unmanned aerial vehicle (UAV), unmanned surface vehicle (USV), and unmanned underwater vehicle (UUV). Drones can be used for fish stock assessment, monitoring and controlling MPAs, and providing evidence for the prosecution of illegal activities. For example, camera-equipped drones can detect fishing vessels and identify if they are registered in a given jurisdiction. They can also observe the use of fishing gears of IUU fishing events and relay real-time the vessel's location and movement to enforcement departments, with the information gathered used to aid in prosecution.[33,34]

Vessel Monitoring Systems

VMS are satellite-based systems that provide data to the fisheries authorities at regular intervals on vessels' location, course, and speed. VMS has become a standard tool globally and provides a greater level of monitoring than conventional aerial and surface technologies. When VMS are installed permanently on a fishing vessel, each unit has a unique identifier used in conjunction with GPS to calculate the vessel's position at routine intervals. States are increasingly entering into multilateral data-sharing agreements that provide "peer-to-peer" VMS information exchanges. VMS is also integrated with other management tools to provide a range of services, as summarised in Table 4.5.[35,36]

Key Elements of Designing and Implementing an Electronic Monitoring Programme

Every fishery is unique in terms of gear types and vessels used, the goals of management, the level of engagement of fishers in the management process, the institutional framework that governs management, and the level of infrastructure in the fishery. As such, there is no "one-size-fits-all" process for establishing an EM programme. Nevertheless, Table 4.6 lists a set of critical elements that can inform the design and implementation of a successful EM programme.[37] Overall, a successfully implemented EM programme can provide a wide range of benefits, including the following:

- *Cost savings*: Particularly in cases when using observers is expensive
- *Employment*: Hiring people to review data and maintain systems

Table 4.5 Services Derived from Integrating Vessel Monitoring Systems with Management Tools

Service	Description
Electronic catch reporting (e-logs)	Facilitate near-real-time catch reporting to allow fishery managers to more easily correlate catch and effort data with VMS position information and inspection reports
Integrated catch documentation schemes	Track and trace fish from the point of capture through the supply chain, essentially tracking from hook to plate, by recording and certifying information that identifies where, when, and by whom the fish were caught
Observer programmes	Onboard observers independently collect information at sea, for example, tracking bycatch, catch composition, and gear configuration data. When coupled and verified with VMS, this information is critical for responsible fisheries management
Catch share or quota monitoring	Catch shares, or quotas, allocate a specific area or percentage of a fishery's total catch to an individual, community, or association. VMS can help hold participants accountable by providing near-real-time information on vessel position as well as catch reporting via e-logs

- *Transparency*: By allowing vessel owners or fishing companies to monitor catches and activities on their vessels to ensure their legality
- *Compliance*: By helping to document conformity with conservation and management measures and international obligations
- *Quality of life at sea*: By reducing the number of observers needed on vessels with limited space
- *Climate resiliency*: By capturing widespread data on fish populations and habitat conditions to better inform adaptive management

Table 4.6 Key Elements of Designing and Implementing an Electronic Monitoring Programme

Element	Outputs/Outcomes	Activities
Motivate EM adoption	Mandates and other incentives are necessary to motivate the investment of time, energy, and resources needed to design and implement an EM programme	• Identify or create a social, regulatory, or statutory commitment to EM • Design communications and conduct meetings to inform and motivate stakeholders
Assemble an EM working group	The EM working group is responsible for designing the EM programme in a participatory way that creates industry support, which is essential for EM programmes to function well	• Identify necessary skills and perspectives • Recruit working group members
Set clear objectives	Fishery management goals must be connected to specific monitoring objectives that guide the development of the EM programme	• Review fishery management objectives • Identify gaps in data streams necessary for achieving objectives
Establish governance for the EM programme	Establish roles and responsibilities	• Roles and responsibilities for every aspect of the EM programme must be made clear and committed to by the responsible parties
Design and optimise the EM programme	Develop an EM system diagram showing the components of the EM system and how they related to each other, and with other monitoring programmes	• Identify EM system components • Articulate how existing monitoring elements will interact with the EM system • Specify infrastructure requirements

(continued)

Table 4.6 (continued)

Element	Outputs/Outcomes	Activities
	Initial specification of EM equipment, data management and rules for data use in making enforcement and management decisions	• Review equipment options and make choices based on capabilities, cost, fit with fishery infrastructure and operations and other considerations • Describe how data will be analysed and used for decision-making • Develop a service model specifying who will oversee the EM system and how EM services will be provided
Understand and articulate the EM value proposition	Perceptions of the costs and benefits of EM compared with those of alternative monitoring programmes often vary within a fishery; a common understanding must be reached to decide whether to develop an EM programme	• Identify costs and benefits relative to other monitoring options
Practical learning through pilots	EM tools and processes should be tested on vessels to prevent problems during implementation	• Install equipment on a few representative vessels • Evaluate equipment performance and data processing; modify as needed
Communication and outreach	Effective two-way communication is essential for engaging all stakeholders in the EM design and implementation process to understand and address concerns	• Identify key audiences and favoured communication modes • Develop key messages • Identify issues that will require workshops

(continued)

Table 4.6 (continued)

Element	Outputs/Outcomes	Activities
Implementation, optimisation, evaluation, and adaptation	Much will be learned during implementation, and conditions will change over time, so evaluation of EM programme performance and periodic adjustment will be required	• Retain vendors and personnel • Install EM equipment • Analyse and apply EM data

- *24/7 coverage*: EM is not affected by differences in working times or weather and is less intrusive than accommodating an extra person onboard
- *Scalability*: Despite the upfront cost, once minimum standards are in place, EM becomes a scalable option for implementation on various gears/vessel types
- *Data integrity*: EM is not susceptible to observer and deployment effects, bribery, intimidation, coercion, or other forms of human bias[38]

Case: Trialling Water-landing Drones Flying Over a Marine Reserve in Belize

In 2019, two drones were used to detect and document illegal fishing activities and conduct ecological research in the Turneffe MPA. The water-landing and waterproof fixed-wing, long-range, multi-camera drones were trialled to monitor and survey marine megafauna, including turtles, dolphins, and sharks, as well as gather evidence of IUU fishing. The drones with their front live-link HD cameras gave conservation officers the ability to spot diving boats immediately. The pair of drones was also operated in various beyond visual line of sight operations (BVLOS) scenarios, with 24 BVLOS flights successfully conducted, with an average length of 10.9 kms and total transect lengths of 263 kms.[39]

Case: Satellite Monitoring of Fishing Boats in Australia
VMS are employed by the Australian Fisheries Management Authority for the delivery of near real-time vessel information via International Maritime Satellite (INMARSAT) communications to monitor the movements of all Commonwealth endorsed fishing vessels effectively. Each VMS unit routinely produces positional reports which contain information such as the vessel's current location, course, and speed for domestic compliance, fisheries management, and research. For instance, the tool assists fisheries managers to achieve compliance, particularly when fishing activities need to be restricted to certain areas or zones. The VMS also provides automatic alert notifications designed to raise awareness when the vessel enters protected areas such as Commonwealth Marine Reserves.[40]

Voluntary Fish Certification Schemes

Voluntary third-party certification for fish and seafood products is a market-based incentive to promote sustainable capture fisheries. It is driven by inadequate fishery management practices, unsustainable depletion of stocks, and ecological degradation. Certification schemes are typically developed by non-governmental organisations and the private sector and supported by consumers who are encouraged to purchase products that have ecolabels through awareness-raising programmes. Schemes usually fall into one of the following categories:

1. *First-party schemes*: These are established by individual companies based on their product standards. The standards may be based on specific environmental issues known to informed consumers through the media or advertising. These schemes are usually referred to as self-declaration
2. *Second-party schemes*: Industry associations establish these for their members' products. The members develop certification criteria that

may be guided by academia and environmental organisations. Verification of compliance is achieved through internal certification procedures within the industry or employment of external certifying companies

3. *Third-party schemes*: These are usually established by an initiator (public or private) independent from the producers, distributors, and sellers of labelled products. Certified products are labelled with information that the product was produced in an environmentally friendly manner. Producers are usually expected to track the "chain of custody" of their products to ensure that the products derived from the certified fishery are those that are so labelled[41]

Where relevant, the costs of pre-assessments, fishery assessment, and periodic re-assessment are usually paid by the producers. Certified fisheries have to undergo annual audits and make improvements accordingly, with re-assessment every few years. At times, governments and other funding sources will subsidise the cost of ecolabelling, particularly for small-scale fisheries.[42]

Case: The Marine Stewardship Council's Blue Fish Label

The Marine Stewardship Council's (MSC) blue fish label is only applied to wild fish or seafood from fisheries that have been certified to the MSC Fisheries Standard, a science-based set of requirements for sustainable fishing. An ecosystem approach is taken with fisheries assessed according to the sustainability of the fish stocks they target, impacts in the broader marine environment, including habitats and other species, and how effectively they are managed. Furthermore, all along the supply chain, MSC certified fish and seafood can be traced through MSC's Chain of Custody Standard to ensure that every distributor, processor, and retailer trading in MSC certified seafood has effective traceability systems in place. To sell MSC labelled seafood, each company along the supply chain must have a valid MSC Chain of Custody certificate and pass regular independent audits to retain it. During each audit, the company needs to demonstrate that:

- The certified seafood it sells originates from a certified seller

- It has systems in place to segregate and prevent mixing between certified and non-certified seafood
- The certified products are identifiable at all times
- A traceability system is in place so that any product sold can be traced back to a certified supplier
- All products can be traced forwards from the point of purchase to the point of sale

The MSC also conducts biannual DNA tests on random samples of MSC labelled products to verify that these traceability requirements are effective.[43,44]

Guiding Principles for Sustainable Fisheries Management

A set of guiding principles can form the basis for developing and implementing sustainable fisheries:

- Fisheries management should make use of the best available scientific information and should aim to achieve both conservation and sustainable use objectives
- Conservation should prioritise the protection, maintenance, and rehabilitation of genetic diversity within both species and ecosystems. The principle should apply to all species, from those of immediate commercial importance to those caught incidentally or negatively impacted by fishing operations
- Sustainable use should be based on pre-determined management plans and respond to changes in stock status. This can be achieved by implementing a harvest strategy framework, incorporating biological reference points (limits and targets), and feedback harvest control rules. Control rules should maintain the stock in the vicinity of the target with high probability while only allowing an exceptionally low probability of stock falling below the limit reference point at any time.

The harvest strategy should ensure that stocks below target reference points rebuild within prescribed timeframes

- Harvest control rules need to be evaluated through simulation testing, preferably before implementation, to determine whether they are likely to perform satisfactorily under a variety of fishery behaviours, change in stock size and productivity, and range of uncertainty
- Conservation and sustainable use involve trade-offs. Management strategies for capture fisheries should be developed in cooperation with stakeholders and groups representative of public interests, for example, conservation and recreation, to fully reflect societal values[45]

Ecosystem Approach to Fisheries

The 1992 UN Convention on Biological Diversity (CBD) refers to the "ecosystem approach". It defines it as ecosystem and natural habitats management that meets "*human requirements to use natural resources, whilst maintaining the biological richness and ecological processes necessary to sustain the composition, structure and function of the habitats or ecosystems concerned. Important within this process is the setting of explicit goals and practices, regularly updated in the light of the results of monitoring and research activities*". It was also defined as "*a strategy for the integrated management of land, water, and living resources that promotes conservation and sustainable use in an equitable way*", as well as "*a strategy…to reach balance between… conservation, sustainable use, and fair and equitable sharing of the benefits arising from the utilisation of genetic resources*".[46]

Traditionally, fisheries management has focused on "single-species management", an approach that does not always recognise the complexity and dynamic nature of marine environments and the inherent levels of uncertainty involved in predicting the outcomes of the interactions with an ecosystem.[47] An ecosystem approach to fisheries (EAF) is required to halt the decline and maintain stocks in the long term. An EAF is an integrated management approach across coastal and marine areas and their natural resources that promotes conservation and sustainable use of the whole ecosystem. Specifically, EAF views a fishery within a wider ecosystem, with management decisions considering the interactions within and between species and between species and their environment.[48] It is a practical way to achieve the principles of sustainable development and has been adopted by the Food and Agricultural Organization Committee on Fisheries as the appropriate approach for the management of fisheries.[49] Implementing EAF involves answering four questions about how a fishery is contributing to sustainable development:

- What impacts are the fishing activities having on target and associated species plus the broader ecosystem?
- What impacts are these fishing activities having on the resources or human activities managed by other sectors?
- What are the economic/social benefits and costs of fishing and related activities to the sector and society as a whole?
- What other activities and drivers beyond fishery management control affect the fishery's capacity to reach its management objectives?[50]

Five vital elements are necessary for implementing EAF, all of which require constant engagement with all marine environment users. These elements are:

- *Data collection and analysis*: EAF requires collecting comprehensive data on the stocks and habits of both target and non-target species within an ecosystem. The good use of available data significantly improves the probability of setting appropriate catch limits. Fishers often hold valuable data, so strong partnerships between the fishing and scientific communities are crucial for acquiring locally held or traditional knowledge. EAF also requires an understanding of the biology and interactions of non-target and target species
- *Precautionary management*: According to the Food and Agricultural Organization Code of Conduct for Responsible Fisheries, the "*absence of adequate scientific information should not be used as a reason for postponing or failing to take measures to conserve target species, associated or dependent species and non-target species and their environment*". As such, under uncertain conditions or a lack of data, management decisions

should be based on the precautionary approach using the best available data to prevent over-exploitation of a fishery. An adaptive management approach enables improvements to fisheries with the sustainability of fisheries able to be improved over time using the best available data
- *Managing competing uses*: Adopting EAF requires the comprehensive collection of marine data as well as requiring that all users of the marine environment, including the fishing industry, aquaculture, energy companies, shipping companies, government interests, and conservation groups are brought together to make coordinated decisions about how to use their marine resource
- *Marine protected areas*: MPAs can provide benefits for fisheries management, including the protection of juvenile commercially relevant fish, a reduction in pressure on stocks during critical breeding periods, and the provision of pristine areas. MPAs can be implemented as the temporary rolling closure of an area or permanent area closure
- *Reducing bycatch*: The reduction of bycatch is essential as the capture of juveniles, females, or species of importance is damaging to a fishery's functionality and ecosystem. Bycatch reduction is crucial to any transition towards a sustainable and resilient fishery. In addition, bycatch reduction provides economic benefits for fishers, including a reduction in catch of non-target species leaving more time and space onboard for target species[51]

Guiding the Implementation of the Ecosystem Approach to Fisheries

The main task of fisheries authorities, with stakeholder participation, is to develop the EAF concept into an operational implementation process at the regional, national, or local level. EAF requires a policy, strategy, and an operational management plan:

1. *Policy*: The policy states the EAF's commitments. It is based on political, scientific, and pragmatic considerations. The policy can be linked to national and local developments and legally binding

or voluntary international agreements. It defines high-level conceptual goals and constraints and connects fisheries management to sector-development planning. It also integrates socio-economic and environmental considerations. High-level goals may include food security, sustainable livelihoods, environmental and resource rehabilitation, economic efficiency that achieves human and ecosystem well-being, maintaining ecosystem integrity, ensuring equity within and between generations, and promoting improved stewardship. The policy also discusses allocation and user rights, clarifying existing and future allocation instruments, as well as conflict management mechanisms

2. *Strategy*: The strategy turns the conceptual goals into operational objectives, ranking them and defining the timeframe they should obtain. The objectives may include reducing the impact on target and non-target species, protecting or rehabilitating habitats and biodiversity, reducing risk to the resource and people, improving food security, or improving governance. For each operational objective, there needs to be an indicator and reference value defined, agreed, and used to determine whether the objective is being achieved. The strategy includes the instruments, measures, technical regulations, and controls needed to achieve them
3. *Management plan*: The management plan details the resources available, the stakeholders involved, the measures specific to the various fisheries, and the enforcement mechanisms utilised. The plan specifies expected outputs and outcomes, allowing overall performance to be evaluated. It identifies indicators and reference values corresponding to the main objectives and constraints through interaction among the main stakeholders. It enables feedback and adaptation as better information is obtained. The process can be guided by a framework of:

    1. Scoping (mapping resources, issues, stakeholders, competing uses, existing rights)
    2. Collecting background information and analysing (ex-ante assessments, strategic analysis, synergies, conflicts)
    3. Setting operational objectives with indicators and reference points

4. Formulating decision rules
5. Implementing and enforcing
6. Monitoring
7. Conducting an ex-post assessment and review[52]

Ecosystem-based Fisheries Management

Ecosystem management involves the direct manipulation of the habitat and population in space and age structure and human activity, with the view to optimising long-term returns to humans. It is area-based, and boundaries need to be clearly and formally defined. It aims to maintain ecosystems in the sustainable condition necessary to achieve desired social benefits.[53] In the United States, the National Marine Fisheries Service defines EBFM as "*a systematic approach to fisheries management in a geographically specified area that contributes to the resilience and sustainability of the ecosystem; recognises the physical, biological, economic, and social interactions among the affected fishery-related components of the ecosystem, including humans; and seeks to optimise benefits among a diverse set of societal goals*".[54,55] The overall goal of EBFM is to sustain healthy marine ecosystems and the fisheries they support. EBFM should:

- Avoid degradation of ecosystems as measured by indicators of environmental quality and system status
- Minimise the risk of irreversible change to natural assemblages of species and ecosystem processes
- Obtain and maintain long-term socio-economic benefits without compromising the ecosystem
- Generate knowledge of ecosystem processes sufficient to understand the likely consequences of human actions[56]

Integrated Ecosystem Assessments

Integrated ecosystem assessments organise science to inform decisions on EBFM.[57] These assessments provide an ability to assess ecosystem status relative to objectives of different groups (for example, fishing, recreation, energy production, shipping, agriculture, forestry, food, clean water), account for the holistic impact of management decisions, and guide management evaluations. Their analyses help resource managers make informed and effective decisions. Typical assessments contain:

- Assessments of status and trends of the ecosystem conditions, including ecosystem services
- Assessments of activities or elements in an ecosystem that can stress it
- Prediction of the future condition of the ecosystem under stress if no management action is taken
- Prediction of the future condition of the stressed ecosystem under different management scenarios
- Evaluation of the success of management actions in achieving the desired target conditions[58]

Implementing Ecosystem-based Fisheries Management

Implementing EBFM is not a single large action but a series of ongoing, cumulative actions that lead to the comprehensive management of living marine resources. The implementation of EBFM can follow six guiding principles with the overall aim of maintaining productive and resilient ecosystems:

1. *Implement ecosystem-level planning*: The planning process should describe and integrate ecosystem goals, objectives, and priorities for fisheries and ecosystem research, conservation, and management across multiple fisheries within an ecosystem
2. *Advance the understanding of ecosystem processes*: Governing authorities should work to understand better the broader suite of ecosystem processes, drivers, threats, status, and trends of their marine ecosystems to inform all levels of management. Greater understanding can come from science projects to understand ecosystems and the compiling of ecosystem status reports for large marine ecosystems

3. *Prioritise vulnerabilities and risks to ecosystems and their components*: Authorities should evaluate and address the individual and cumulative drivers for the physical, chemical, biological, social, and economic components of marine ecosystems. The process should consider the comprehensive and systematic risk, vulnerability, and susceptibility of living marine resources and ecosystems, including:
   - Identifying the ecosystem-level, cumulative risk (across living marine resources, habitats, ecosystem functions, and associated fisheries communities) in regions and the relative vulnerability to human and natural pressures
   - Identify the individual and cumulative pressures that pose the most risk to those vulnerable resources and dependent communities
4. *Explore and address trade-offs within an ecosystem*: In close cooperation with stakeholders, authorities should consider various trade-offs when considering the cumulative effects of decision-making processes on the ecosystem, including:
   - Analysing trade-offs to optimise total benefits from all fisheries within each ecosystem or jurisdiction by considering regional socio-economic considerations and ecosystem-specific policy goals and objectives
   - Developing management strategy evaluation capabilities to conduct better ecosystem-level analyses that provide ecosystem-wide management advice
5. *Incorporate ecosystem considerations into management advice*: Living marine resources management should consider best available ecosystem science in decision-making processes, including developing and monitoring ecosystem-level reference points, incorporating ecosystem considerations into living marine resources assessments, rules, and management decisions, and providing integrated advice for other management decisions, particularly those decisions applied across multiple species within an ecosystem
6. *Maintain resilient ecosystems*: Authorities should aim to sustain resilient and productive living marine resources populations and habitats, maintain overall ecosystem structure and function, and support

the contributions that fisheries make to the socio-economic resiliency of coastal human communities. EBFM should develop operating protocols to maintain resilient ecosystems. Actions in support of this include:

- Evaluating ecosystem-level measures of resilience to maintain core ecosystem structure, biodiversity, production, energy flow, and functioning
- Evaluating coastal fishing community well-being[59][60]

Case: Bering Sea Fishery Ecosystem Plan

In 2018, the North Pacific Fishery Management Council adopted the Bering Sea Fishery Ecosystem Plan (BS FEP). The BS FEP establishes a framework for the Council's progress towards EBFM of the Bering Sea fisheries and describes how it will support research projects to address the Council's priorities. The Council has adopted five prioritised Action Modules for the BS FEP. These Action Modules are projects that turn the BS FEP into real-life changes:

1. *Evaluate short- and long-term effects of climate change on fish and fisheries and develop management considerations*: The goal is to evaluate the vulnerability of critical species and fisheries to climate change and strengthen resilience in regional fisheries management. Specifically, this Action Module will provide a seven-year climate context to interpret and respond to annual signals and establish a more formal process for considering those variables. This is responsive to the BS FEP purpose of building resiliency into the Council's management strategies and providing options for responding to changing circumstances
2. *Develop protocols for using Local Knowledge (LK) and Traditional Knowledge (TK) in management and understanding impacts of Council decisions on subsistence use*: This Action Module has two parts. In Part A, integrating/incorporating LK and TK into Council processes in the short- to long-term will be addressed. In Part B, a methodology will be developed for how the Council can consider potential impacts

to subsistence species, habitats that support those species, and access to subsistence resources

3. *Gap analysis of Bering Sea management with EBFM best practices*: This assessment will serve as an internal assessment of the Council's state of EBFM practice and a gap analysis of areas where there may be an opportunity for further action. Such a gap analysis will help to prioritise areas of future work for Council management and other Action Modules
4. *Interdisciplinary conceptual models for the Bering Sea ecosystem*: This involves developing non-quantitative interdisciplinary "conceptual models" (system diagrams) of the Bering Sea ecosystem through an interdisciplinary process to highlight key ecosystem components and detail conceptual understandings of pressures and drivers that contribute to the status and trends, including habitat areas of particular concern. This will help the Council assess trade-offs of management actions on different components of the ecosystem, leading to more informed decision-making
5. *Align and track Council priorities with research funding opportunities*: This Action Module focuses on tracking research relevant to the BS FEP Action Modules and how that information is subsequently used in management. It will enable the Council to identify more effective ways to engage with researchers addressing the Council's research priorities to ensure that results are relevant and valuable in management[61,62]

Notes

1. FAO, "Achieving Blue Growth: Building Vibrant Fisheries and Aquaculture Communities," (2018), http://www.fao.org/policy-support/tools-and-publications/resources-details/en/c/1234976/.
2. The Prince's Charities' International Sustainability Unit, "Towards Global Sustainable Fisheries: The Opportunity for Transition," (2012), https://www.greengrowthknowledge.org/sites/default/files/

downloads/resource/Towards_global_sustainable_fisheries_Prince%27s%20Charities.pdf.
3. FAO, "Fisheries Management: 4. Marine Protected Areas and Fisheries," (2011), http://www.fao.org/3/i2090e/i2090e.pdf.
4. The Prince's Charities' International Sustainability Unit, "Towards Global Sustainable Fisheries: The Opportunity for Transition."
5. World Bank, "Giving Oceans a Break Could Generate US$83 Billion in Additional Benefits for Fisheries," https://www.worldbank.org/en/news/press-release/2017/02/14/giving-oceans-a-break-could-generate-83-billion-in-additional-benefits-for-fisheries.
6. John T. Trochta et al., "Ecosystem-Based Fisheries Management: Perception on Definitions, Implementations, and Aspirations," *PLOS ONE* 13, no. 1 (2018).
7. OECD, "Encouraging Policy Change for Sustainable and Resilient Fisheries," (2019), https://www.oecd-ilibrary.org/agriculture-and-food/encouraging-policy-change-for-sustainable-and-resilient-fisheries_31f15060-en.
8. The Prince's Charities' International Sustainability Unit, "Towards Global Sustainable Fisheries: The Opportunity for Transition."
9. Low Impact Fisheries of Europe, "Rights Based Management and Small Scale Fisheries in the Eu: Human Rights Versus Property Rights," (2016), https://lifeplatform.eu/wp-content/uploads/2017/01/LIFE-Statement-on-ITQs_Abridged-version.pdf.
10. FAO, http://www.fao.org/fishery/topic/16200/en.
11. Jamie C. Afflerbach et al., "A Global Survey of "Turf-Reserves," Territorial Use Rights for Fisheries Coupled with Marine Reserves," *Global Ecology and Conservation* 2 (2014).
12. FAO, "Measuring and Assessing Capacity in Fisheries: 1. Basic Concepts and Management Options," (2004), http://www.fao.org/3/y5442e/y5442e00.htm#Contents.
13. "Fao Technical Guidelines for Responsible Fisheries - Fisheries Management - 4," (1999), http://www.fao.org/3/w4230e00.htm#Contents.
14. NOAA, "National Bycatch Reduction Strategy," (2016), https://repository.library.noaa.gov/view/noaa/17062.

15. R. Lewison et al., "Understanding Impacts of Fisheries Bycatch on Marine Megafauna," *Trends in Ecology & Evolution* 19, no. 11 (2004).
16. Ministry for Primary Industries, "Fish Quota Management System," https://www.mpi.govt.nz/legal/legal-overviews-legislation-standards/fisheries-legislation/quota-management-system/.
17. Jean-Yves Weigel et al., "Marine Protected Areas and Fisheries: Bridging the Divide," *Aquatic Conservation: Marine and Freshwater Ecosystems* 24, no. S2 (2014).
18. Sylvaine Giakoumi et al., "Revisiting "Success" and "Failure" of Marine Protected Areas: A Conservation Scientist Perspective," *Frontiers in Marine Science* 5 (2018).
19. FAO, "Fisheries Management: 4. Marine Protected Areas and Fisheries."
20. Ecomagazine, "Ivory Coast Creates First Marine Protected Area," https://www.ecomagazine.com/news/policy/ivory-coast-creates-first-marine-protected-area
21. IPBES, "Locally Managed Marine Area," https://ipbes.net/policy-support/tools-instruments/locally-managed-marine-area#:~:text=A%20Locally%20Managed%20Marine%20Area,or%20are%20based%20in%20the.
22. H. Govan, Aalbersberg, W., Tawake, A., and Parks, J., "Locally-Managed Marine Areas: A Guide for Practitioners," (2008), https://www.reefresilience.org/wp-content/uploads/LMMA-Guide-2008.pdf.
23. Stacy D. Jupiter et al., "Locally-Managed Marine Areas: Multiple Objectives and Diverse Strategies," *Pacific Conservation Biology* 20, no. 2 (2014).
24. Fiji Shores and Marinas, "Flmma Fiji Locally Managed Marine Protected Area Network," https://lmmanetwork.org/who-we-are/country-networks/fiji/.
25. Locally-Managed Marine Area (LMMA) Network, "Lmma Fiji".
26. Environmental Defense Fund, "Designing and Implementing Electronic Monitoring Systems for Fisheries," (2018), http://fisherysolutionscenter.edf.org/sites/catchshares.edf.org/files/EM_DesignManual_Final_0.pdf.

27. Fisheries New Zealand, "Monitoring and Observing," https://www.mpi.govt.nz/protection-and-response/sustainable-fisheries/strengthening-fisheries-management/monitoring-and-observing/.
28. FAO, "Guidelines for Developing an at-Sea Fishery Observer Programme," (2003), http://www.fao.org/3/Y4390E/y4390e00.htm#Contents.
29. "Catch Documentation Schemes: Practices and Applicability in Combating Iuu Fishing," http://www.fao.org/in-action/globefish/fishery-information/resource-detail/en/c/426994/.
30. "Expert Consultation on Catch Documentation Schemes (Cds)," (2015), http://www.fao.org/3/i5063e/i5063e.pdf.
31. European Commission, "Catch – Information Note," (2021), https://ec.europa.eu/fisheries/sites/fisheries/files/docs/body/catch-it-system_en.pdf.
32. The Pew Charitable Trusts, "Electronic Monitoring: A Key Tool for Global Fisheries," (2019), https://www.pewtrusts.org/en/research-and-analysis/issue-briefs/2019/09/electronic-monitoring-a-key-tool-for-global-fisheries.
33. OECD, "An Inventory of New Technologies in Fisheries," (2017).
34. Hilde M. Toonen and Simon R. Bush, "The Digital Frontiers of Fisheries Governance: Fish Attraction Devices, Drones and Satellites," *Journal of Environmental Policy & Planning* 22, no. 1 (2020).
35. European Commission, "Vessel Monitoring System (Vms)," https://ec.europa.eu/fisheries/cfp/control/technologies/vms_en.
36. The Pew Charitable Trusts, "Tracking Fishing Vessels around the Globe," (2017), https://www.pewtrusts.org/en/research-and-analysis/issue-briefs/2017/04/tracking-fishing-vessels-around-the-globe.
37. Environmental Defense Fund, "Designing and Implementing Electronic Monitoring Systems for Fisheries."
38. The Pew Charitable Trusts, "Electronic Monitoring: A Key Tool for Global Fisheries."
39. Aeromao, "Aeromapper Talon Amphibious Fly Bvlos Missions over Marine Reserve for Illegal Fishing Detection and Biodiversity Research," https://www.aeromao.com/2019/08/02/aeromapper-talon-amphibious-routinely-fly-bvlos-missions-over-marine-reserve-for-illegal-fishing-detection-and-biodiversity-research/.

40. Australian Fisheries Management Authority, "Satellite Monitoring of Fishing Boats," https://www.afma.gov.au/monitoring-enforcement/satellite-monitoring-fishing-boats.
41. FAO, "Product Certification and Ecolabelling for Fisheries Sustainability," (2001), http://www.fao.org/tempref/docrep/fao/005/y2789e/y2789e00.pdf.
42. "Seafood Certification and Developing Countries: Focus on Asia," (2018), http://www.fao.org/3/i8018en/I8018EN.pdf.
43. MSC, "What Does the Blue Msc Label Mean?" https://www.msc.org/what-we-are-doing/our-approach/what-does-the-blue-msc-label-mean.
44. "From Ocean to Plate: How DNA Testing Helps to Ensure Traceable, Sustainable Seafood," (2016), https://www.msc.org/docs/default-source/default-document-library/what-we-are-doing/msc-from-ocean-to-plate-traceability-and-dna-report-2016.pdf?sfvrsn=fbce98a4_4.
45. Peter A. Shelton and Alan F. Sinclair, "It's Time to Sharpen Our Definition of Sustainable Fisheries Management," *Canadian Journal of Fisheries and Aquatic Sciences* 65, no. 10 (2008).
46. FAO, "The Ecosystem Approach to Fisheries: Issues, Terminology, Principles, Institutional Foundations, Implementation and Outlook," (2003), http://www.fao.org/3/a-y4773e.pdf.
47. Andrew J. Kenny et al., "Delivering Sustainable Fisheries through Adoption of a Risk-Based Framework as Part of an Ecosystem Approach to Fisheries Management," *Marine Policy* 93 (2018).
48. The Prince's Charities' International Sustainability Unit, "Towards Global Sustainable Fisheries: The Opportunity for Transition."
49. FAO, "What Is Eaf," http://www.fao.org/fishery/eaf-net/about/what-is-eaf/en.
50. Ibid.
51. The Prince's Charities' International Sustainability Unit, "Towards Global Sustainable Fisheries: The Opportunity for Transition."
52. Serge M. Garcia and Kevern L. Cochrane, "Ecosystem Approach to Fisheries: A Review of Implementation Guidelines," *ICES Journal of Marine Science* 62, no. 3 (2005).

53. FAO, "The Ecosystem Approach to Fisheries: Issues, Terminology, Principles, Institutional Foundations, Implementation and Outlook."
54. NOAA, "Ecosystem-Based Fisheries Management Policy," (2018), https://www.fisheries.noaa.gov/resource/document/ecosystem-based-fisheries-management-policy.
55. Kenny et al., "Delivering Sustainable Fisheries through Adoption of a Risk-Based Framework as Part of an Ecosystem Approach to Fisheries Management."
56. E. K. Pikitch et al., "Ecosystem-Based Fishery Management," *Science* 305, no. 5682 (2004).
57. Christian Möllmann et al., "Implementing Ecosystem-Based Fisheries Management: From Single-Species to Integrated Ecosystem Assessment and Advice for Baltic Sea Fish Stocks," *ICES Journal of Marine Science* 71, no. 5 (2014).
58. NOAA, "Understanding Ecosystem-Based Fisheries Management," https://www.fisheries.noaa.gov/insight/understanding-ecosystem-based-fisheries-management#:~:text=Ecosystem%2Dbased%20fisheries%20management%20is,species%20or%20issue%20in%20isolation.
59. "Ecosystem-Based Fisheries Management Policy."
60. Ibid.
61. North Pacific Fishery Management Council, "Bering Sea Fishery Ecosystem Plan," https://www.npfmc.org/bering-sea-fishery-ecosystem-plan/.
62. "Bering Sea Fishery Ecosystem Plan," (2019), https://meetings.npfmc.org/CommentReview/DownloadFile?p=c334ad33-4139-4b5a-b205-a8b7c5028562.pdf&fileName=D6%20Final%20BS%20FEP%20Jan%202019.pdf.

References

Aeromao. "Aeromapper Talon Amphibious Fly Bvlos Missions over Marine Reserve for Illegal Fishing Detection and Biodiversity Research." https://www.aeromao.com/2019/08/02/aeromapper-talon-amphibious-routinely-fly-bvlos-missions-over-marine-reserve-for-illegal-fishing-detection-and-biodiversity-research/.

Afflerbach, Jamie C., Sarah E. Lester, Dawn T. Dougherty, and Sarah E. Poon. "A Global Survey of "Turf-Reserves", Territorial Use Rights for Fisheries Coupled with Marine Reserves." *Global Ecology and Conservation* 2 (2014/12/01/ 2014): 97–106.

Australian Fisheries Management Authority. "Satellite Monitoring of Fishing Boats." https://www.afma.gov.au/monitoring-enforcement/satellite-monitoring-fishing-boats.

Ecomagazine. "Ivory Coast Creates First Marine Protected Area." https://www.ecomagazine.com/news/policy/ivory-coast-creates-first-marine-protected-area.

Environmental Defense Fund. "Designing and Implementing Electronic Monitoring Systems for Fisheries." (2018). http://fisherysolutionscenter.edf.org/sites/catchshares.edf.org/files/EM_DesignManual_Final_0.pdf.

European Commission. "Catch – Information Note." (2021). https://ec.europa.eu/fisheries/sites/fisheries/files/docs/body/catch-it-system_en.pdf.

———. "Vessel Monitoring System (Vms)." https://ec.europa.eu/fisheries/cfp/control/technologies/vms_en.

FAO. http://www.fao.org/fishery/topic/16200/en.

———. "Achieving Blue Growth: Building Vibrant Fisheries and Aquaculture Communities." (2018). http://www.fao.org/policy-support/tools-and-publications/resources-details/en/c/1234976/.

———. "Catch Documentation Schemes: Practices and Applicability in Combating Iuu Fishing." http://www.fao.org/in-action/globefish/fishery-information/resource-detail/en/c/426994/.

———. "The Ecosystem Approach to Fisheries: Issues, Terminology, Principles, Institutional Foundations, Implementation and Outlook." (2003). http://www.fao.org/3/a-y4773e.pdf.

———. "Expert Consultation on Catch Documentation Schemes (Cds)." (2015). http://www.fao.org/3/i5063e/i5063e.pdf.

———. "Fao Technical Guidelines for Responsible Fisheries - Fisheries Management - 4." (1999). http://www.fao.org/3/w4230e00.htm#Contents.

———. "Fisheries Management: 4. Marine Protected Areas and Fisheries." (2011). http://www.fao.org/3/i2090e/i2090e.pdf.

———. "Guidelines for Developing an at-Sea Fishery Observer Programme." (2003). http://www.fao.org/3/Y4390E/y4390e00.htm#Contents.

———. "Measuring and Assessing Capacity in Fisheries: 1. Basic Concepts and Management Options." (2004). http://www.fao.org/3/y5442e/y5442e00.htm#Contents.

———. "Product Certification and Ecolabelling for Fisheries Sustainability." (2001). http://www.fao.org/tempref/docrep/fao/005/y2789e/y2789e00.pdf.

———. "Seafood Certification and Developing Countries: Focus on Asia." (2018). http://www.fao.org/3/i8018en/I8018EN.pdf.

———. "What Is Eaf." http://www.fao.org/fishery/eaf-net/about/what-is-eaf/en.

Fiji Shores and Marinas. "Flmma Fiji Locally Managed Marine Protected Area Network." https://lmmanetwork.org/who-we-are/country-networks/fiji/.

Fisheries New Zealand. "Monitoring and Observing." https://www.mpi.govt.nz/protection-and-response/sustainable-fisheries/strengthening-fisheries-management/monitoring-and-observing/.

Garcia, Serge M., and Kevern L. Cochrane. "Ecosystem Approach to Fisheries: A Review of Implementation Guidelines." *ICES Journal of Marine Science* 62, no. 3 (2005): 311–18.

Giakoumi, Sylvaine, Jennifer McGowan, Morena Mills, Maria Beger, Rodrigo H. Bustamante, Anthony Charles, Patrick Christie, et al. "Revisiting "Success" and "Failure" of Marine Protected Areas: A Conservation Scientist Perspective." *Frontiers in Marine Science* 5 (2018).

Govan, H., Aalbersberg, W., Tawake, A., and Parks, J. "Locally-Managed Marine Areas: A Guide for Practitioners." (2008). https://www.reefresilience.org/wp-content/uploads/LMMA-Guide-2008.pdf.

IPBES. "Locally Managed Marine Area." https://ipbes.net/policy-support/tools-instruments/locally-managed-marine-area#:~:text=A%20Locally%20Managed%20Marine%20Area,or%20are%20based%20in%20the.

Jupiter, Stacy D., Philippa J. Cohen, Rebecca Weeks, Alifereti Tawake, and Hugh Govan. "Locally-Managed Marine Areas: Multiple Objectives and Diverse Strategies." *Pacific Conservation Biology* 20, no. 2 (2014): 165–79.

Kenny, Andrew J., Neil Campbell, Mariano Koen-Alonso, Pierre Pepin, and Daniela Diz. "Delivering Sustainable Fisheries through Adoption of a Risk-Based Framework as Part of an Ecosystem Approach to Fisheries Management." *Marine Policy* 93 (2018): 232–40.

Lewison, R., L. Crowder, A. Read, and S. Freeman. "Understanding Impacts of Fisheries Bycatch on Marine Megafauna." *Trends in Ecology & Evolution* 19, no. 11 (2004): 598–604.

Locally-Managed Marine Area (LMMA) Network. "Lmma Fiji." Locally-Managed Marine Area (LMMA) Network.

Low Impact Fisheries of Europe. "Rights Based Management and Small Scale Fisheries in the Eu: Human Rights Versus Property Rights." (2016). https://lifeplatform.eu/wp-content/uploads/2017/01/LIFE-Statement-on-ITQs_Abridged-version.pdf.

Ministry for Primary Industries. "Fish Quota Management System." https://www.mpi.govt.nz/legal/legal-overviews-legislation-standards/fisheries-legislation/quota-management-system/.

Möllmann, Christian, Martin Lindegren, Thorsten Blenckner, Lena Bergström, Michele Casini, Rabea Diekmann, Juha Flinkman, et al. "Implementing Ecosystem-Based Fisheries Management: From Single-Species to Integrated Ecosystem Assessment and Advice for Baltic Sea Fish Stocks." *ICES Journal of Marine Science* 71, no. 5 (2014): 1187–97.

MSC. "From Ocean to Plate: How DNA Testing Helps to Ensure Traceable, Sustainable Seafood." (2016). https://www.msc.org/docs/default-source/default-document-library/what-we-are-doing/msc-from-ocean-to-plate-traceability-and-dna-report-2016.pdf?sfvrsn=fbce98a4_4.

———. "What Does the Blue Msc Label Mean?" https://www.msc.org/what-we-are-doing/our-approach/what-does-the-blue-msc-label-mean.

NOAA. "Ecosystem-Based Fisheries Management Policy." (2018). https://www.fisheries.noaa.gov/resource/document/ecosystem-based-fisheries-management-policy.

———. "National Bycatch Reduction Strategy." (2016). https://repository.library.noaa.gov/view/noaa/17062.

———. "Understanding Ecosystem-Based Fisheries Management." https://www.fisheries.noaa.gov/insight/understanding-ecosystem-based-fisheries-management#:~:text=Ecosystem%2Dbased%20fisheries%20management%20is,species%20or%20issue%20in%20isolation.

North Pacific Fishery Management Council. "Bering Sea Fishery Ecosystem Plan." https://www.npfmc.org/bering-sea-fishery-ecosystem-plan/.

———. "Bering Sea Fishery Ecosystem Plan." (2019). https://meetings.npfmc.org/CommentReview/DownloadFile?p=c334ad33-4139-4b5a-b205-a8b7c5028562.pdf&fileName=D6%20Final%20BS%20FEP%20Jan%202019.pdf.

OECD. "Encouraging Policy Change for Sustainable and Resilient Fisheries." (2019). https://www.oecd-ilibrary.org/agriculture-and-food/encouraging-policy-change-for-sustainable-and-resilient-fisheries_31f15060-en.

———. "An Inventory of New Technologies in Fisheries." (2017).

Pikitch, E. K., C. Santora, E. A. Babcock, A. Bakun, R. Bonfil, D. O. Conover, P. Dayton, *et al.* "Ecosystem-Based Fishery Management." *Science* 305, no. 5682 (2004): 346–47.

Shelton, Peter A., and Alan F. Sinclair. "It's Time to Sharpen Our Definition of Sustainable Fisheries Management." *Canadian Journal of Fisheries and Aquatic Sciences* 65, no. 10 (2008/10/01/ 2008): 2305–14.

The Pew Charitable Trusts. "Electronic Monitoring: A Key Tool for Global Fisheries." (2019). https://www.pewtrusts.org/en/research-and-analysis/issue-briefs/2019/09/electronic-monitoring-a-key-tool-for-global-fisheries.

———. "Tracking Fishing Vessels around the Globe." (2017). https://www.pewtrusts.org/en/research-and-analysis/issue-briefs/2017/04/tracking-fishing-vessels-around-the-globe.

The Prince's Charities' International Sustainability Unit. "Towards Global Sustainable Fisheries: The Opportunity for Transition." (2012). https://www.greengrowthknowledge.org/sites/default/files/downloads/resource/Towards_global_sustainable_fisheries_Prince%27s%20Charities.pdf.

Toonen, Hilde M., and Simon R. Bush. "The Digital Frontiers of Fisheries Governance: Fish Attraction Devices, Drones and Satellites." *Journal of Environmental Policy & Planning* 22, no. 1 (2020/01/02/ 2020): 125–37.

Trochta, John T., Maite Pons, Merrill B. Rudd, Melissa Krigbaum, Alexander Tanz, and Ray Hilborn. "Ecosystem-Based Fisheries Management: Perception on Definitions, Implementations, and Aspirations." *PLOS ONE* 13, no. 1 (2018): e0190467.

Weigel, Jean-Yves, Kathryn Olivia Mannle, Nathan James Bennett, Eleanor Carter, Lena Westlund, Valerie Burgener, Zachary Hoffman, *et al.* "Marine Protected Areas and Fisheries: Bridging the Divide." *Aquatic Conservation: Marine and Freshwater Ecosystems* 24, no. S2 (2014): 199–215.

World Bank. "Giving Oceans a Break Could Generate US$83 Billion in Additional Benefits for Fisheries." https://www.worldbank.org/en/news/press-release/2017/02/14/giving-oceans-a-break-could-generate-83-billion-in-additional-benefits-for-fisheries.

5

Sustainable Aquaculture

Introduction

Aquaculture is defined by the Food and Agricultural Organization as the "*farming of aquatic organisms, including fish, molluscs, crustaceans and aquatic plants. Farming implies some form of intervention in the rearing process to enhance production, such as regular stocking, feeding, protection from predators,* etc. *Farming also implies individual or corporate ownership of the stock being cultivated*". Aquaculture in brackish and marine waters, also known as "mariculture", is where aquatic organisms—both plants and animals—are cultured in a confined environment in the aquatic medium, which may be wholly marine or marine mixed to various degrees with freshwater in the brackish water areas.[1] This chapter will first discuss the types of aquaculture and systems and types of structures used in brackish and marine aquaculture operations before discussing monitoring methods. The chapter will then discuss the concepts of sustainable aquaculture, integrated multi-trophic aquaculture, and organic aquaculture before reviewing spatial planning for aquaculture. Finally, the chapter will examine the ecosystem approach to aquaculture.

R. C. Brears, *Developing the Blue Economy*,
https://doi.org/10.1007/978-3-030-84216-1_5

Types of Aquaculture and Systems

There are four types of aquaculture by the degree of commercialisation:

- Aquaculture for subsistence (family-level)
- Artisanal aquaculture, producing for the market on a small-scale
- Specialised aquaculture in which different farmers carry out various stages of the production cycle
- Industrial-scale aquaculture

The types of aquaculture mentioned above can be compared with the standard classification of aquaculture based on productive technology, mainly feed, dividing culture systems into the following:

- *Extensive culture systems*: These systems receive no intentional nutritional inputs but depend on natural food in the culture facility, including that brought in by water flow (for example, currents and tidal exchange)
- *Semi-intensive culture systems*: These systems depend primarily on natural food, which is increased over baseline levels by fertilisation and/or use of supplementary feed to complement natural food
- *Intensive culture systems*: These systems depend on nutritionally complete diets added to the system, either fresh, wild, marine, or freshwater fish, or on formulated diets, usually in dry pelleted form[2]

Types of Aquaculture Structures

The main types of structures used in aquaculture operations in brackish and marine environments are:

1. *Floating cages*: Floating cages are utilised in various farm sizes and installed in the open sea or large sheltered bays. The open exchange of water through the nets replenishes oxygen and removes dissolved and solid wastes. Because of the high costs of cages, it is necessary to maintain a larger number of individuals inside the cage to ensure the

venture is profitable. In turn, this necessitates supplementary feeding as the quantity of natural food available in the volume of water inside the cage cannot sustain the large number normally cultured in these cages

2. *Net enclosures*: Net enclosures that barricade off large areas in sheltered bays are being implemented, but it requires significant capital outlay and frequent replacement because of corrosion. In addition, supplementary feed is essential, increasing running costs
3. *Coastal ponds*: Coastal ponds and lagoons have been used for centuries for fish, mollusc, crustacean, and seaweed production. Oysters and mussels are mainly grown on structures above the seabed and feed on microscopic algae floating in the water. Sites used for coastal ponds are tidal and intertidal mudflats in protected areas of estuaries, bays, creeks, lagoons, and marshes
4. *Constant water circulation systems*: Constant water circulation systems are large cement structures. The systems require the continuous pumping of water in large quantities and supplementary feed[3,4,56]

Waste

Rapid population growth and increased demand for fish has resulted in the intensification of aquaculture, increasing waste generation from production systems. Source of waste from aquaculture include:

- *Feed*: In extensive systems, feed supply to fish is mainly dependent on the environment. In semi-intensive systems, fish are stocked in moderate to high density and rely on food from both natural production and supplemental feed from the culturists, while in intensive systems, high-quality artificial feed is used to achieve fast growth
- *Chemicals*: Some chemicals are used, including various pharmaceutical products, to treat and control parasites and microbial infections. Antifouling paints and food are also potential sources of chemical contaminants

- *Pathogens*: Natural water bodies have their pathogenic load, and receiving additional loads from fish culture systems may cause stress or the outright death of aquatic organisms

Components of Waste

The components of waste produced from aquaculture systems include:

- *Solid wastes*: Solid wastes are mainly derived from the uneaten feed and faecal droppings of cultured fish. They sometimes include fish that do not survive the culture process. Solid waste can be further classified as suspended solids and settled solids. In addition to solid wastes being dangerous as they can clog fish gills and lead to death, their aerobic bacterial activity will increase the chemical oxygen demand and biochemical oxygen demand and deplete oxygen in the culture column
- *Dissolved wastes*: Dissolved wastes are products of food metabolism in fish or decomposed uneaten feed. In dissolved waters, the two main components of concern are nitrogen and phosphorous products. These two elements are the main component of fish feed. When these nutrients are released into the water, they can harm fish and other inhabitants of aquatic ecosystems at high concentrations. Nitrogen is released into culture water in the form of ammonia, which is highly toxic to both fish cultured in the system and those in receiving water bodies, while nitrate, which is the end product of ammonia oxidation, can accumulate over time, enriching the receiving water, and with phosphorous, cause eutrophication[7]

Management of Waste

The primary solution for managing the environmental impacts of aquaculture is the management of feed. Feed and feeding systems can effectively reduce wastes resulting from the fish feed through proper management of the inputs into the culture system. The following is recommended to reduce feed-related waste from aquaculture:

- The species and fish size-specific potential performance of any diet to be supplied must be known. The labelling of feed can include information on feed digestibility and waste production, including the quantity of solids, phosphorous, and nitrogen
- There should be knowledge of the biomass of the fish in the system
- Adequate information on the health and physiological status of the fish must be available
- Uniformity in size of fish is important, allowing the fish to accept the same size of pellet
- The feed should be sieved to remove dust and broken pellets before being fed
- The feed must be fed effectively to ensure little or no waste[8,9]

Monitoring

When farms have been established, it is essential to monitor them for environmental impacts.[10] A tiered approach is recommended for monitoring all forms of aquaculture. Different levels of monitoring are required according to the type of aquaculture and the farm's history in terms of management and environmental performance. The approach is adaptive and encourages efficient monitoring and sustainable management practices. The levels of monitoring are detailed in Table 5.1.[11]

Monitoring Non-feed Aquaculture

It is recommended that a two-tiered approach is taken when monitoring non-feed aquaculture farms. A relatively low level of seabed monitoring (Level 1) is recommended when either the intensity of farming is considered not to be on an industrial scale, when farming is only seasonal, or where farms have been in place for many years the effects are well-documented. A greater intensity of monitoring (Level 2) is recommended where a farm is to be developed in a new area or region and where the effects are not well-documented, which involves:

Table 5.1 Levels of Aquaculture Monitoring

Level	Description
Level 1	• The least intense form of monitoring • It places greater emphasis on qualitative indicator variables that can be rapidly evaluated and reported within weeks • It focuses on assessment at two to three monitoring stations, including one or two located at the outer limit of effects, and one at a near-field reference location
Level 2	• Involves monitoring at two or three stations within the zone of maximum effects, one or two stations at the outer limit of effects, and at near-field and far-field reference locations • Multiple replicate samples of total free sulphides, redox potential, sediment texture, total organic content, total nitrogen, and total phosphorus are taken
Level 3	• The most intensive type of monitoring • It aims to detect spatial patterns or address specific concerns • It is conducted at year 0 (baseline) and after a few years before being conducted when required • The aim is to gain a better understanding of the causal factors and develop an adaptive management plan response • Two types of Level 3 sampling are: - Sampling regularly along radial transects to review whether the monitoring captures the zone of maximum effect - Sampling over a grid pattern to map the distribution and extent of habitats and resulting footprint (e.g., a pre-farm baseline or after five years to cross-check actual against the predicted footprint

1. *Sediment physicochemical characteristics*: Sediment samples are collected at a representative number of sites beneath areas being farmed and at reference locations. Samples can be collected using a remotely controlled grab or core sampler or by divers using SCUBA. Sediment samples can be analysed for total free sulphides, redox potential, sediment texture, total organic content, total nitrogen, and total phosphorus
2. *Infauna species and abundance*: A sediment infauna core can be collected from the same locations used to assess the sediment's physicochemical characteristics. The organisms within the core are sampled
3. *Epifauna species and abundance*: Conspicuous epibiota are quantitatively assessed from randomly position photo-quadrants around each

monitoring site. Benthic photo-quadrant images are taken with a high-resolution camera. Images are analysed, and conspicuous biological features such as bacterial mats and burrows are identified. The density of shellfish on the seabed is also enumerated from these photo-quadrants

4. *Video/photographic and visual information*: Video footage and/or standardised photographs can be used to assess general habitat, shellfish densities, evidence of sediment bioturbation, algal and bacteria mat development and so forth[12]

Monitoring Feed-added Aquaculture

A three-tiered approach for monitoring is recommended for aquaculture farms involving the addition of feed. The level of monitoring differs between farms. Level 1 monitoring is recommended where the effects of a farm are well-documented and have historically stayed within acceptable limits. Level 3 monitoring is recommended for new farms involving high-level feed and/or managing a farm at the upper limits of environmental thresholds. Progression to less intensive monitoring—from Level 2 to 1—is dependent on how long the farm has been operational, whether feed levels have increased significantly, and whether the results of the previous years' annual monitoring are compliant with rules and regulations.[13]

Case: Research on Smart Buoys to Monitor Mussel Farms
New Zealand has around 645 green shell mussel farms, with the most prominent being over 20 hectares. The farms are made up of numerous longlines that support crop lines that the mussels attach themselves to. Keeping the longlines afloat are buoys, with 50–70 buoys per longline. Spin-off research from the Precision Farming Technologies for Aquaculture Spearhead Project, within the Science for Technological Innovation National Science Challenge, aims to turn these buoys into real-time sensors that monitor the state of the farms. The buoys, equipped with Internet of Things (IoT) sensors, aim to provide each farmer with the

exact conditions at individual mussel farms. The sensors could be used to monitor water quality and water turbidity, temperature, and ocean acidity, which mussel larvae are sensitive to. A low-power IoT network connects the sensors in a network and communicates the data back to shore.[14]

Sustainable Aquaculture

Because aquaculture is practised by both some of the most impoverished farmers in developing countries and by multinational companies, sustainable aquaculture strategies need to be devised to ensure:

- Farmers receive a fair income from farming
- The benefits and costs are shared equitably
- The promotion of wealth and job creation
- Enough food is accessible for all
- The environment is managed for the benefit of future generations
- That aquaculture development is orderly, with both authorities and industry well organised[15]

A set of guiding principles are recommended that can guide the development of sustainable aquaculture to ensure it is developed to its full potential so that communities prosper and people are healthier, there are increased opportunities for improved livelihoods, with increased income and better nutrition, and farmers and women are empowered:

- *Adherence to Rio Principles*: Aquaculture should adhere to the Rio principles of sustainable development, particularly intergenerational equity, the precautionary approach, and the polluter pays principle:
  - *Intergenerational equity*: At the core of sustainable development is the emphasis on intergenerational equity. Aquaculture may change the balance and distribution of different resources or capital,

such as natural capital, human capital, physical plant (equipment, machinery, buildings) and infrastructure, financial capital, and other forms of capital valued by society. These changes need to be assessed to ensure that the sum of this capital, or its specific vital components, are sustained or increased in the long-run, and are available for future generations
  - *Precautionary approach*: The precautionary approach implies that we should carefully plan and rigorously evaluate developments that have uncertain implications for the environment. This approach ensures that planning takes into complete account both the magnitude and likelihood of adverse environmental impacts, and so there is some form of risk assessment that takes place
  - *Polluter pays principle*: This principle ranges from requirements to pay for clean-up costs to the responsibility to pay for the cost of environmental damage as well as clean-up costs

- *Integration and coordination*: Aquaculture planning should involve various stakeholders, with differing values, in the planning process. Aquaculture planning should be coordinated between different sector policies. Furthermore, planning should consider, or be coordinated with, location initiatives and regional or national level policies. Nonetheless, increased integration is likely to slow down and make it more difficult for the decision-making process. Therefore, there are a variety of strategies that can be implemented to reduce associated complexity, including:
  - Developing high quality, well presented, and effectively communicated/exchanged information
  - Creating a clear and widely agreed decision criteria
  - Ensuring clear and transparent decision-making processes
  - Creating a clearly designated and widely agreed, final authority and judge (whether individual or committee)

- *Aquaculture should involve the public in decision-making processes*: Significant public involvement is a desirable and necessary part of any planning initiative for many reasons, including involving local people can provide essential information about local natural resources, their status, use, and value; public involvement can reduce conflict through the early identification and resolution of potentially contentious issues; a more participatory form of public involvement can allow otherwise under-represented groups access to the decision-making process. It can take a variety of forms, including:

  - *Communication of information*: From decision-makers, planners, or technical specialists to other stakeholders, and vice versa
  - *Participation*: There should be shared responsibility and decision-making

- *Assessment of costs and benefits*: Aquaculture management often targets output maximisation rather than profit maximation. In addition to this being economically inefficient, it can also lead to increased social and environmental risks. Sustainable aquaculture can be efficient and sustainable when the costs and benefits are assessed objectively or comprehensively. Aquaculture decisions should be based on thorough assessments of costs and benefits (financial, economic, social, environmental) of aquaculture in a specific area and a comparative assessment done of the costs and benefits of aquaculture relative to other resource uses
- *Estimation of environmental capacity*: Environmental capacity assessments, which is, in this context, the magnitude of aquaculture production that can be supported without leading to significant changes to ecological processes, species, populations, or communities in the environment, can be used both as a tool to support decisions, for example, when measured alongside a more extensive Environmental Impact Assessment, and as a process and guiding approach to making decisions about development, for example, to set limits on production
- *Emphasis on incentives rather than constraints*: Incentives rather than regulations should guide the development of the sector. Regulatory approaches to planning and management are often complex, costly, and unpopular, and at times need streamlining in highly regulated countries with slow maritime growth. In contrast, incentives stimulate innovation, leading to more environmentally friendly technologies. The use of economic instruments can influence both the siting and operation of aquaculture operations

- *Control of effects rather than the scale of activity*: Many forms of regulation of aquaculture are related to scale, usually in terms of water area directly used by aquaculture or the total production, for example, an upper limit on aquaculture production in a bay or estuary. This limits economic development and provides no incentive to improve the environmental efficiency of the operation. A limit on effects, for example, the concentration of nitrogen in the water at critical times of the year, would provide an incentive for improved environmental efficiency through technology or management
- *Evaluation, iteration, and adaptation*: Management decisions should involve repeated cycles of evaluation, iteration, and adaptation to allow for the steady refinement and improved understanding of the physical, ecological, social, and economic parameters and processes of sustainable aquaculture over time and allow for the steady refinement and improvement of the planning instruments (incentives and constraints) used to meet sustainable aquaculture objectives[16,17,18,19,20,21]

Case: Sustainable Aquacultural Production in Spain

Veta La Palma is a 113 square kilometre estate located in the Doñana Natural Reserve of Spain. PIMSA has been running an extensive and semi-intensive aquaculture project on 32 square kilometres there since the 1990s. Veta La Palma produces 800–1,000 tonnes of seafood per annum (sea bass, bream, grey mullet, and shrimp) in large, naturally flooded ponds in the coastal wetlands. A low density of a maximum of four kilograms of live animal per square metre reduces the pressure on the surrounding environment while guaranteeing high animal welfare. Aquaculture production is integrated with dry agriculture, rice production, and natural wetlands. Specifically, the production area comprises 45 ponds, each covering 173 acres interconnected with the Guadalquivir and Guadiamar rivers by a complex network of irrigation and drainage channels that total nearly 300 kms in length. The system is connected to a station with a capacity to pump up to 12,000 L/second through open, semi-open, or closed systems as required by environmental and crop conditions. The ponds are used for extensive breeding of sea bass, bream, grey mullet, and shrimp, while smaller ponds at the head of each of the larger ones are used for decantation, confinement, and fattening of young fish and breeding on a semi-intensive basis. The aquaculture project attracts around 250 bird species, half of which are threatened in the area. The initiative has also restored previously degraded marshlands, which are now again water ecosystems, allowing the recovery of the associated biodiversity and ecosystem services.[22,23]

Integrated Multi-trophic Aquaculture

Integrated multi-trophic aquaculture (IMTA) is a practice where uneaten feed and waste of one species are recaptured and converted into feed, fertilisers, and energy for another species. In contrast to polycultures, where several species can be cultured together, but all belong to the same trophic level, IMTAs include species from different trophic levels, for instance: the co-cultivation of fed species (such as finfish or shrimp); together with extractive species, such as suspension-feeding (for example, mussels and oysters) and deposit-feeding invertebrates (for example, sea cucumbers and sea urchins) as well as macroalgae (for example, kelp, which may feed on the organic and inorganic effluents generated by the fed species). The specific components of IMTA farming systems are listed in Table 5.2. By creating integrated cultivation systems, the sustainability of aquaculture is increased as IMTA allows the creation of more sustainable production systems because wastes of fish/shrimp production are valued as a resource instead of being considered pollution. At the same time, IMTA increases economic resilience from increased production efficiency, product diversification, and potential price premiums, with each level having its independent commercial value. Finally, IMTA can teach students and inform the public about environmental sustainability, increase their ocean literacy, and improve the social acceptance of aquaculture products.[24,25,26,27,28]

Table 5.2 Components of Integrated Multi-trophic Aquaculture Systems

Component	Description
Fish (the fed component)	Some farmed species require manufactured feeds with portions going uneaten. The wastes produced by fish, which include uneaten feed and faeces, provide high-quality nourishment for other species within the IMTA system, including wild species
Filter feeders (the organic fine particulate extractive component)	Filter-feeding bivalves, such as mussels and oysters, filter the water column feeding on microalgae, small zooplankton, and fine particulate matter. They can be used to reduce the level of finer organic particles that result from other fed or non-fed components of the IMTA system
Deposit (or bottom) feeders (the organic large particulate nutrients extractive component)	Under an IMTA model, these are primarily invertebrates, such as sea cucumbers, sea urchins, and certain worm species that sift through sediment to feed on organic particulate matter. They can be used to recycle the larger organic particles that result from the other (fed or non-fed) components of the IMTA system and that settle beneath the farm site
Seaweeds (the inorganic dissolved nutrient extractive component)	Kelps and other seaweeds naturally extract dissolved inorganic nutrients, for example, nitrogen and phosphorous. They can help reduce the levels of dissolved inorganic nutrients generated by the other fed and non-fed components of the IMTA system. The seaweed component of the IMTA system is usually placed further away to capture better the inorganic dissolved nutrients that are lighter and travel longer distances than the organic nutrients

Species Selection

Environmental sustainability is the primary consideration of IMTA. The criteria guiding species selection are the imitation of the natural ecosystem. The combinations of co-cultured species are carefully selected according to a few conditions and criteria, including the following:

- *Complementary role with other species in the system*: Species are used that complement each other on different trophic levels. For instance, species must be able to feed on the other species' waste for the newly integrated species to improve the quality of water and grow efficiently

- *Adaptability in relation to the habitat*: Native species within their normal geographic range and for which technology is available can be used. This prevents the risk of invasive species causing harm to the local environment and potentially impacting other economic activities
- *Ability to provide both efficient and continuous bio-mitigation*: Species should be able to grow to significant biomass. This is important if the organisms act as a biofilter that captures many of the excess nutrients and can be harvested from the water. The alternative is to have species with a very high value, and so lesser volumes can be grown
- *Market demand for the species and pricing as raw material for their derived products*: Use species with an established or perceived market value. Farmers must be able to sell the alternative species to increase their economic output
- *Commercialisation potential*: Use species that regulators and policymakers will help develop new markets for, rather than impose regulations on their commercialisation
- *Environmental performance*: Ensure choices made contribute to improved environmental performance
- *Social and political issues*: Ensure choices made are compatible with a variety of social and political issues[29]

Case: All Atlantic Ocean Sustainable, Profitable and Resilient Aquaculture Project

The European Union (EU)-funded All Atlantic Ocean Sustainable, Profitable and Resilient Aquaculture (ASTRAL) project will develop IMTA production chains for the Atlantic markets. The project is led by the Norwegian research centre NORCE with consortium partners from Scotland, Ireland, France, Spain, Portugal, Nigeria, South Africa, Argentina, and Brazil. The overall objective of ASTRAL is to develop new, sustainable, profitable, and resilient value chains for IMTA production within the framework of existing, emerging, and potential Atlantic markets. The specific objectives of ASTRAL are listed in Table 5.3.[30,31]

Table 5.3 Objectives of the ASTRAL Project

Objective	Activity
Validate cost-effective IMTA processes	Develop and innovate techniques and species combinations to validate cost-effective IMTA processes from a regional perspective, including fish, mollusc, echinoderm, crustacean, and algae species
Explore local species	Explore local species from the Beagle Channel in Argentina for IMTA production, including fish, crustacean, mollusc, earthworms, and algae species
Provide a profitable IMTA production system	Provide farmers of aquatic organisms a profitable system, bringing diversification and increasing profits by at least 30%
Create a catalogue of Atlantic species for the future	Develop IMTA production manuals and best practices to implement IMTA throughout the Atlantic regions
Develop new and improved technologies	Develop new and improved technologies, including bivalve sensors, real-time monitoring IoT kits, Artificial Intelligence (AI)-based vision sensors, an improved imaging-based microplastic sensor, and an AI data analytics platform
Circularity	Increase circularity by 50–60% through IMTA production, compared to monoculture baselines and determine the best configurations towards zero waste
Complying with organic standards	Assess IMTA requirements to comply with organic standards
Assess impacts of IMTA labs	Assess the environmental and social impacts of IMTA labs in Ireland, South Africa, Scotland, and Brazil
Foster collaboration	Provide a collaborative ecosystem for understanding and harnessing the Atlantic Ocean resources sustainably
Capacity development	Improve professional skills and create a highly trained workforce
Identify microplastics	Identify and characterise potential environmental discharges of micron-sized plastic fragments within the IMTA recirculating inshore system

(continued)

Table 5.3 (continued)

Objective	Activity
Identify potential environmental and climatic risks	Identify potential environmental and climatic risks for Northern and Southern Atlantic regions and mitigate risks related to these threats
Provide economic and business tools to aquaculture producers	Provide economic and business tools for IMTA adoption and implementation by delivering new business models for IMTA production and a cost–benefit analysis
Share knowledge, technology, and practices	Share co-generated knowledge, technology, practices, and efforts with and for society to encourage pro-poor and gender-sensitive development
Cross-governance assessments	Carry out cross-Atlantic governance assessments and IMTA value chain consumer perceptions and social licence assessments
Disseminate activities and result visibly	Disseminate and ensure visibility of ASTRAL activities and results at international events, high-level and technical meetings, workshops and conferences, and articles in academic journals

Organic Aquaculture

Organic aquaculture is the farming of aquatic animals, such as shrimp, fishes, bivalves, and aquatic plants, without using antibiotics, chemicals, and fertilisers, which preserves the ecosystem and biodiversity. Organic aquaculture production aims to minimise its environmental impact as much as possible while developing a valuable and sustainable aquatic ecosystem.[32] Organic aquaculture involves raising aquatic products in a humane manner that is sustainable and pollution-free. For example, good health management practices minimise stress by providing free movement, appropriate living conditions, and optimum stocking relative to the carrying capacity of the farming system. In addition, organic feed must be from certified sustainable sources, and harvested products should be pollution-free and sustainably produced. The nutrient

Table 5.4 Differences Between Conventional and Organic Aquaculture

Conventional Aquaculture	Organic Aquaculture
Intensive farming system, high stocking density	Extensive farming system, low stocking density, and encouraging polyculture
Results in environmental degradation	Complete environmental monitoring network and environmental management system to protect and improve functions and processes of the natural ecosystems
Produces chemical and nutrient pollution, contaminating territorial and coastal waters	Minimisation of chemical or nutrient pollution
Results in net protein consumption as well as increased pressure on wild resources	Results in net protein production
Involves limited recycling of nutrients and directly discharges waste into the surrounding environment	Nutrient recycling and regeneration
Does not always use native or resident aquatic animal species	Uses native or resident aquatic animal species
Uses chemical materials and antibiotics etc	Prohibits the use of synthetic chemical materials, antibiotics, etc
Often abandons non-productive production bases without rehabilitation	Takes an ecosystem rehabilitation and enhancement approach
Limited farm lifespans and abandonment	Long-term production plan for sustainable development
No consideration of the impact of production on the surrounding environment	Keeps farming within the carrying capacity of the local environment and is consistent with local environmental programmes

content of fertilisers is varied and includes cattle dung, biogas slurry, compost, poultry droppings, and oil cakes. The main differences between conventional and organic aquaculture are summarised in Table 5.4.[33,34]

Organic Aquaculture Standards

In 2002, the International Federation of Organic Agriculture Movements published the first draft of basic standards for organic aquaculture, which became entirely accepted basic standards five years later. The standards are:

- Absence of genetically modified organisms in stocks and presence of vegetable feed ingredients (e.g., soybeans)
- Limitation of stocking density; considering the ecological capacity of the site and species-specific behaviour of animals
- Use of vegetable feed and fertiliser from certified organic agriculture which promotes recycling of nutrients instead of intensive input
- No use of synthetic pesticides and herbicides to maintain natural diversity on the farm area
- Restriction of energy consumption to facilitate de-intensification of operations and lowering of input
- Preference must be given to natural medicines and the absence of prophylactic use of antibiotics and chemotherapeutics
- Intensive monitoring of environmental impact, protection of the surrounding ecosystem, and integration of natural plant communities in farm management focusing on the waste management
- Processing should be according to organic principles, and the final products certified as organic

Organic Aquaculture Certification

Organic aquaculture certification programmes contain standards related to environmental protection, social responsibility, and food safety. Aquaculture labelling can be integrated into existing certifications or done as part of a stand-alone labelling scheme. Standards for organic aquaculture certification are continuously modified to include new ideas and to comply with international guidelines. The basic steps towards organic certification are:

1. *Exchange of information*: The first step is the exchange of information. This provides detailed information about technical and formal aspects of certification. A basic questionnaire survey is conducted in this step
2. *Pre-evaluation visit*: The purpose of this visit is to get an impression of the situation on-site, discuss the steps towards conversion with all

parties involved, and set up a conversion plan. The visit also helps assist the farm/organisation in preparing for the upcoming inspection
3. *Inspection*: Before the inspection is scheduled, a cost estimation is issued by the inspection body. Following the inspection, the inspector will issue the inspection report to the certificate agency for listing and evaluating of the findings
4. *Contracts*: Assuming a positive decision by the certification committee, a contractual partnership between a farm and the standard-setting organisation will be established
5. *Certification*: The inspection report and further data and information are forward to the certification committee. The committee's decision is communicated to the farm by the certification letter containing the conditions to which the certification is subjected. It also decides on the annual renewal of certification[35,36]

Case: Europe's Organic Logo Identifying Organic Aquaculture Products

The EU organic logo was launched in 2010 to provide a visual identity to EU-produced organic products, helping consumers identify organic products and helping farmers market them across the entire EU. The organic logo can only be used on products that have been certified as organic by an authorised control agency or body. This means they have met strict conditions on how they are produced, processed, transported, and stored. The logo can only be used on products with at least 95% organic ingredients and meet strict conditions for the remaining five percent. EU organic farming labelling rules encompass every stage of the production process, with specific rules governing the organic aquaculture sector. The main features of the organic aquaculture labelling rules include:

- Strict maximum stocking densities
- Water quality requirements
- Rules that specific that biodiversity should be respected, and which do not allow the use of induced spawning by artificial hormones
- Handling minimised to avoid stress and physical damage

- The provision that organic feeds should be used, supplemented by fish feeds derived from sustainably managed fisheries
- Special provisions are made for bivalve mollusc production and seaweed[37]

Spatial Planning for Aquaculture

In the marine environment, there are many users, including aquaculture, tourism, fisheries, marine transport and so on, all of which have different objectives, goals, and resource needs, often putting them in direct conflict with each other. Therefore, selecting the spatial area designated for aquaculture development and the careful selection of farm sites is the first step to ensure the success and sustainability of aquaculture.

Spatial planning for aquaculture should consider the environmental, economic, social, and governance objectives of sustainable development, especially when aquaculture occurs in shared water resources. Overall, the concept of sustainability should be embedded in aquaculture planning and management to ensure that aquaculture development is ecologically sustainable and enables the rational use of resources shared by aquaculture and other activities.[38,39] The process of spatial planning usually consists of three steps: First, aquaculture zoning, second, site selection, and third, the design of aquaculture management areas (AMAs).

Aquaculture Zoning

The selection of the spatial area designated for aquaculture development usually occurs at two levels as follows.

National-level Scoping

Scoping helps governments proactively and strategically plan for sustainable aquaculture development and management. The main objective is to define the boundaries of management units and the ecosystem, determine the relative importance of development and conservation goals, and ensure stakeholders are well-informed about the costs and benefits of aquaculture development. The process is usually led by an aquaculture task force that consults with relevant stakeholders, including government officials, policymakers, scientists, farmers, fishers, and other competing marine environment users, to ensure a balanced and successful planning process.[40]

Regional-level Zoning

Zoning is the process of identifying the desired geographic location and extent of aquaculture and other activities in a region based on ecological and socio-economic criteria. Policymakers, government officials, scientists, farmer groups, industry representatives, local authorities and regulatory bodies, and community members should be directly involved in this process. Zoning should aim to minimise adverse environmental impacts, biosecurity risk, and stakeholder conflict. Within the zoning process, stakeholders should ensure that they properly identify suitable areas for aquaculture using the criteria listed in Table 5.5. Once zones have been selected, planners need to determine the carrying capacity of these zones. Carrying capacity in aquaculture is the upper bounds of production that can be sustained based on the available resources. There are four main types of carrying capacity:

- *Physical carrying capacity*: Quantifies the total area in a waterbody suitable for aquaculture. It is often expressed as the number of farms or as the amount of geographic area occupied
- *Production carrying capacity*: Determines the limits of aquaculture production at the farm level, specifically, the stocking density at which production biomass is maximised
- *Ecological carrying capacity*: Estimates the amount of production that can be sustained without causing irreversible damage to or altering ecological processes, species, populations, and habitat
- *Social carrying capacity*: Estimates the amount of aquaculture production that can be supported without generating user conflicts[41,42]

Table 5.5 Zoning Considerations for Aquaculture

Zoning Considerations	Description
Ecological	Suitable zones should have abundant marine waters of adequate quality for target species. In addition, planners should consider how aquaculture's impact on the water column, benthic environment, and surrounding sensitive ecological areas and populations might impact other users (e.g., wild-caught fisheries, tourism) when selecting areas
Socio-economic	Aquaculture is ideally placed in areas with few existing users (e.g., shipping, tourism, wild-caught fisheries) to minimise potential user conflicts, and areas with access to production infrastructure (e.g., roads, energy) and markets for both inputs and outputs
Risks and Issues	Planners need to be aware of issues and risks in all steps of the aquaculture production process, as well as their respective impacts, scales, and the likelihood of occurrence

Site Selection for Individual Aquaculture Farms

Site selection determines where farms will be located, what kind of aquaculture will be permitted, the species to be farmed, and the likely impacts of such proposed aquaculture. The goal is to ensure that sites are appropriately located to maximise production and minimise adverse social and environmental impacts. The private sector typically leads siting, but the government provides structure and standards for site licensing, zonal usage, and environmental impact assessment.[43] The critical steps in the site selection process are:

- *Assessment of suitability for aquaculture*: The assessment typically includes a review of local conditions (for example, temperature, water quantity), historical conditions (such as historical climate data from the local meteorological agency), and some predictive impacts from aquaculture activity and measures to be undertaken to mitigate those impacts. The choice of the site should also aim to take into consideration the location and distance of sensitive habitats, tourist facilities, sites of cultural importance, and other service infrastructure, with a consideration of the potential to impact these activities or be impacted by these activities

- *Detailed estimation of carrying capacity for sites*: Assessments of carrying capacity at the site level are more developed than the assessment at the zonal scale. In most cases, site-level carrying capacity models estimate nutrient inputs to the environment and assess the impacts of sediments on the water column. Often, models assess these impacts against minimum environmental quality standards, often defined nationally through scientific research and set specifically by regulators, who then set a maximum production level. Some models may also assess profitability to ensure the ecological limits defined are profitable for the farmer as well
- *Biosecurity planning and disease control*: Diseases cause around 40% of all losses in aquacultural systems, and so biosecurity is an essential component of proper farm management at the site level. The essential components of a farm- or site-level biosecurity plan are:
  - *Screening and quarantine*: All animals coming into the farm should be certified disease-free and tested for disease on arrival and be quarantined for a period of time to ensure they are not infected
  - *Isolation*: Nets, tanks, and other equipment should be routinely disinfected, and farm workers should maintain good hygiene
  - *Proper handling*: Fish should be kept in well-oxygenated water at an optimum temperature and handled as little as possible
  - *Proper stocking density*: High-density conditions increase the frequency of contact between individual fish, increasing rates of disease transmission and infection
  - *Regular monitoring*: The loss of appetite is the first sign of disease. Fish should be monitored closely during routine feeding to ensure they are eating well and are healthy
  - *Veterinary services*: A veterinarian should sample the farmed stock at regular intervals to ensure any issues are detected early
- *Authorisation arrangements*: Each separate company or legal entity operating within an aquaculture zone will typically require an aquaculture licence or permit that defines species to be cultured; maximum permitted annual production or peak biomass; culture method; site

marking for navigation safety; and any special conditions such as regular environmental surveys and other monitoring[44]

Aquaculture Management Area

AMAs are a collection of farmers and producers that participate in common management practices. While farmers are responsible for the operation and performance of their farms, AMAs establish and implement common management goals and objectives for the betterment of all farms in the area. AMAs typically focus on issues that can be resolved collectively, such as user rights conflicts, limited access to inputs, and management of risk, waste, and disease. AMAs develop management plans that set goals and objectives, common management practices, monitoring programmes, and biosecurity strategies. The main steps in the definition and management of AMAs are:

- *Delineation of management area boundaries with appropriate stakeholder consultation*: Within a defined aquaculture zone, AMA boundaries can be based on biophysical, environmental, socio-economic, and/or governance-based criteria that create one geographical area with an identifiable physical/ecosystem base. For ease of regulation, AMAs should ideally be within one governance administrative unit (e.g., municipal, state, district, region)
- *Establishing an area management entity involving local communities as appropriate*: Where possible, all operating farms within an AMA should be members of a farmers' or producers' association as a means to allow representation in an area management entity, and which can set and enforce among members the norms of responsible behaviour, including, for example, the development of codes of conduct
- *Carrying capacity and environmental monitoring of AMAs*: Estimates of the environmental carrying capacity of the area should be made, and regular surveys conducted to reassess the area. Carrying capacity at the AMA scale could be undertaken, for example, using depositional models (particle tracking) that predict the particulate outputs from

fish cage aquaculture and that can be used in the local-scale assessment of the effects of fish cages on the organic footprint impact on the sediment and sensitive demersal flora and fauna

- *Disease control in AMAs*: Farmers should be encouraged or possibly mandated to follow sound biosecurity practices that provide the framework for disease management on the farm and that are implemented through documented standard operating procedures. At the farm level, the owner or operator should be responsible for ensuring the implementation of biosecurity. Auditing and certification of the efficacy of a biosecurity programme can be provided by a veterinarian and competent government officer
- *Best management practices*: Best management practices (BMPs) are guidelines that promote improved farming practices to increase production through responsible and sustainable aquaculture. BMPs brings the standard of practice of every participating farmer up to a specified acceptable level and is based on science and experience, reflecting the industry's desire to remain at the forefront of good practice
- *Group certification*: The ability to provide third-party auditing and certification through an effective and justifiable biosecurity plan, when applied at the farm or compartment level, can allow farmers to access markets that require disease-status assurances that may not be available on a national level
- *Essential steps in implementing, monitoring, and evaluating a management plan for an AMA*: Implementing a management plan should be time-bound. Two aspects are significant relative to a time frame. The first is to decide on a base year for the management system. This will represent a year (or period) against which progress can be measured. The second time aspect relates to target years or periods by which various aspects of the work plan can be achieved or by which any quantitative programme output should be attained. Overall, it is likely that the management system should span a 5–10 year time frame, but the system will need periodic reviews over shorter time scales during this time. The management plan must address all the relevant issues, have clear and achievable operational objectives for

each issue, and a clear timeline for completion with targets and indicators. Furthermore, the management plan must have responsible people/institutions/entities, adequate funding for each management approach, and have resources to implement the measures as appropriate[45,46]

Data Requirements for the Spatial Planning Process

Overall, the data requirements for the spatial planning process steps are listed in Table 5.6.[47]

Case: Government of Western Australia's Aquaculture Development Zones

The Government of Western Australia is establishing aquaculture development zones as part of its commitment to developing a sustainable aquaculture industry. An aquaculture development zone is a designated area of water selected for its suitability for a specific aquaculture sector, for example, shellfish. The zones provide "investment ready" areas of water with strategic environmental approvals and management policies already in place, allowing commercial operations to be set up without the need for time-consuming, complex, and expensive approval processes. Three aquaculture development zones have been established in Western Australia: one in the Kimberley, one in the Mid-West, and one in Oyster Harbour, Albany. The government has also invested $1.3 million to identify and establish aquaculture development zones on the south coast of Western Australia.[48]

Ecosystem Approach to Aquaculture

The ecosystem approach to aquaculture (EAA) is a strategy for integrating aquaculture within the wider ecosystem such that it promotes sustainable development, equity, and resilience of interlinked socio-ecological systems.[49] Three strategic principles guide the EAA:

Table 5.6 Data Requirements for the Spatial Planning Process

Steps		Tasks	Data Requirements
1.Aquaculture zoning	Scoping	• Collect baseline information on current aquaculture production, markets, and regulatory frameworks • Define national priorities for aquaculture • Set broad objectives • Identify relevant stakeholders to consult	• Economic or market (international and national) feasibility information • Current regulations or institutions relevant to aquaculture development • Aquaculture production, area, and location • Suitability requirements for target culture species
	Zoning	• Identify suitable aquaculture areas • Identify regional issues or threats • Estimate zonal carrying capacity • Develop biosecurity and zoning strategies • Designate zones for aquaculture	• Water quantity and quality • Hydrodynamics and bathymetry • Suitability requirements for target culture species • Accessibility (infrastructure, markets, roads, labour) • Proximity to sensitive habitats, pollution sources, and other fishing and aquaculture zones

(continued)

Table 5.6 (continued)

Steps	Tasks	Data Requirements
2.Site selection	• Assess aquaculture suitability • Estimate site carrying capacity • Plan for biosecurity and disease control • Develop authorisation procedures for proposed sites	• Water quantity and quality • Hydrodynamics and bathymetry • Site suitability and carrying capacity estimates • Accessibility (infrastructure, markets, roads, electricity, inputs) • Proximity to sensitive habitats, pollution sources, and other fishing and aquaculture zones
3.Aquaculture Management Area	• Consult with stakeholders to delineate management area boundaries • Develop a management body and enforce a plan • Establish carrying capacity and environmental and disease monitoring procedures for management areas	• Proximity to nearby farms • Information on: - Waterbody - Water source - Species farmed • Environmental impact information (water turnover, feed conversion rate, benthic diversity, bottom anoxia) • Carrying capacity

1. Aquaculture should be developed in the context of ecosystem functions and services (including biodiversity) with no degradation of these beyond their resilience
2. Aquaculture should improve human well-being with equity, for example, access rights and a fair share of income, for all relevant shareholders
3. Aquaculture should be developed in the context of other sectors, policies, and goals as appropriate[50]

The EAA builds on these guiding principles to provide a planning and management framework for effectively integrating the aquaculture sector into local planning. In addition, the EAA offers an appropriate framework to develop management plans for AMAs that go beyond the individual farms. The enacting of the EAA consists of the following steps:

1. Scoping to understand the broader issues in the multi-stakeholder context in which aquaculture may develop
2. Identification of opportunities and assessment of main risks with special consideration to fish disease and environmental issues
3. Carrying capacity estimation to determine maximum production allowed in a given area
4. Allocation of user/area access and/or management rights
5. Development of management plans for the zone/site/AMA
6. Monitoring of the plan and adjustment over time[51]

Benefits of the Ecosystem Approach to Aquaculture

The enaction of the EAA can provide a wide range of environmental, economic, and social benefits, as summarised in Table 5.7.[52]

Allocated Zones for Aquaculture

With the EAA approach, ecological carrying capacity is essential for ensuring the long-term sustainability of aquaculture operations. If the ecological carrying capacity is exceeded, it can result in a range of

Table 5.7 Benefits of the Ecosystem Approach to Aquaculture

Challenge of Aquaculture	Benefits of the Ecosystem Approach to Aquaculture
Fish disease and lack of effective biosecurity (e.g., when farms are too close to each other)	• Minimise fish disease risk and better response to outbreaks
Environmental issues (e.g., eutrophication and biodiversity and ecosystem service losses, etc.)	• Better coordinated and integrated approaches to the use and management of natural resources • A better understanding of cumulative and combined environmental effects and interactions between users and the environment
Production issues (e.g., lower growth and biomass of filter feeders due to excessive farming density and overharvesting of common pool microalgae)	• Improved filter-feeder productivity and yields
Social conflict (e.g., equity issues and lack of public confidence in the sustainability of aquaculture)	• Improved accountability and transparency through relevant stakeholder involvement at all levels
Post-harvest and marketing issues (e.g., when individual neighbour farmers do not have access to postharvest services)	• Clusters of farmers having better access to common postharvest processes and other services • Area-based management and certification as a governance and risk-sharing model for sustainable aquaculture
Risk financing (National governments and financing institutions do not have a good knowledge of where the prospects for aquaculture development are most promising before committing resources to development)	• National-level information on areas available to invest in aquaculture • Implementing area-based management strategies (e.g., clusters of farmers) to facilitate access to finance
Lack of resilience to climatic variability, climate change and other external threats and disasters (e.g., hurricanes, tsunamis, drought, and industrial pollution of water sources)	• A more resilient sector, better adapted to shocks • More effective mechanisms for governments and other institutions, including civil society organisations, to deliver services and fulfil their commitments to sustainable aquaculture development

negative environmental impacts, including eutrophication, hypoxia, and harmful algal blooms, affecting aquaculture production and other industries. One tool available to improve water quality is the designation of allocated zones for aquaculture (AZA), which are specific areas where aquaculture development is prioritised over other users. AZAs are effective at managing water quality, particularly in areas where other cumulative uses impact water quality, such as industrial discharges and microbial loading in wastewater. However, an AZA does not in itself mandate limits on farms or stocking densities. This can be overcome by establishing allowable zones of effect (AZE) for fish cages in coastal systems. An AZE can help determine the minimum separation of cages, mitigating organic enrichment and hypoxia in bottom sediments.[53]

Case: Determining the Ecological Appropriateness of Proposed Aquaculture Zones in South Australia

The Government of South Australia's Primary Industries and Regions SA Fisheries and Aquaculture (PIRSA FA) develops the State's aquaculture zone policies in consultation with the industry, key state and local government agencies, and the wider community. Aquaculture zone policies are statutory policies that dedicate or prioritise spatial areas of the marine environment for the purposes of aquaculture activity, which includes species that may be farmed, areas where marine aquaculture cannot occur, the type of aquaculture system that is allowed, and limitations on biomass or leasable areas in a given location. Before developing zone policies, PIRSA FA conducts desktop research of the available environmental, conservation, heritage, industrial/commercial, and social data of each area to determine the spatial scope for aquaculture proposed zones. Following this, an independent scientific-technical investigation is undertaken to determine the environmental conditions and sustainable carrying capacity of the region identified. It includes benthic video analysis, water and sediment chemical analysis, sediment infauna analysis, oceanography analysis, and carrying capacity modelling. This provides the technical information needed to determine the ecological appropriateness of the proposed aquaculture zone areas.[54]

Notes

1. FAO, "Mariculture as a Method of Using Coastal Zones," http://www.fao.org/3/AC583E/AC583E00.htm#ch2.
2. "Rural Aquaculture: Overview and Framework for Country Reviews," (1998), http://www.fao.org/3/x6941e/x6941e04.htm#bm04.2.
3. "Mariculture as a Method of Using Coastal Zones."
4. CTA Brussels Office, "Fish-Farming: The New Driver of the Blue Economy," (2013), https://europa.eu/capacity4dev/file/15656/download?token=J7_MU3_d.
5. FAO, "Mariculture as a Method of Using Coastal Zones."
6. CTA Brussels Office, "Fish-Farming: The New Driver of the Blue Economy."
7. Akeem Babatunde Dauda et al., "Waste Production in Aquaculture: Sources, Components and Managements in Different Culture Systems," *Aquaculture and Fisheries* 4, no. 3 (2019).
8. Ibid.
9. Abdolsamad K. Amirkolaie, "Reduction in the Environmental Impact of Waste Discharged by Fish Farms through Feed and Feeding," *Reviews in Aquaculture* 3, no. 1 (2011).
10. FAO, "Sustainable Development of Marine Aquaculture Off-the-Coast and Offshore – a Review of Environmental and Ecosystem Issues and Future Needs in Temperate Zones," (2013), http://www.fao.org/tempref/FI/CDrom/P24/i3530e/root/10.pdf.
11. Waikato Regional Council, "Monitoring Framework for the Waikato Coastal Marine Area: Report 3 – Seabed and Water Column Monitoring and Standards," (2015), https://www.waikatoregion.govt.nz/services/publications/tr201540/.
12. Ibid.
13. Ibid.
14. National Science Challenges, "Let's Turn Those Buoys into Real-Time Mussel Sensors," https://www.sftichallenge.govt.nz/news/lets-turn-those-buoys-real-time-mussel-sensors/.
15. FAO, "Aquaculture," http://www.fao.org/aquaculture/en/.

16. "Planning and Management for Sustainable Coastal Aquaculture," (2001), http://www.fao.org/3/a-y1818e.pdf.
17. C. Byron and Barry Costa-Pierce, "Carrying Capacity Tools for Use in the Implementation of an Ecosystems Approach to Aquaculture," *FAO Fisheries and Aquaculture Proceedings* 21 (2013).
18. Jenny Weitzman and Ramón Filgueira, "The Evolution and Application of Carrying Capacity in Aquaculture: Towards a Research Agenda," *Reviews in Aquaculture* (2019).
19. Stella Tsani and Phoebe Koundouri, "A Methodological Note for the Development of Integrated Aquaculture Production Models," *Frontiers in Marine Science* 4, no. 406 (2018).
20. Christopher M. Free et al., "Realistic Fisheries Management Reforms Could Mitigate the Impacts of Climate Change in Most Countries," *PLOS ONE* 15, no. 3 (2020).
21. FAO, "Aquaculture."
22. Veta la Palma, "Extensive Aquaculture," https://www.vetalapalma.es/en/extensive-aquaculture/.
23. Meta, "Future Farming: Fishing for a Sustainable Seafood Model in Spain," https://meta.eeb.org/2020/07/23/future-farming-fishing-for-a-sustainable-seafood-model-in-spain/.
24. João Rosa et al., "Integrated Multitrophic Aquaculture Systems – Potential Risks for Food Safety," *Trends in Food Science & Technology* 96 (2020).
25. Bela H. Buck et al., "State of the Art and Challenges for Offshore Integrated Multi-Trophic Aquaculture (Imta)," *Frontiers in Marine Science* 5, no. 165 (2018).
26. Marta Correia et al., "Integrated Multi-Trophic Aquaculture: A Laboratory and Hands-on Experimental Activity to Promote Environmental Sustainability Awareness and Value of Aquaculture Products," ibid.7, no. 156 (2020).
27. Fisheries and Oceans Canada, "Aquaculture in Canada: Integrated Multi-Trophic Aquaculture (Imta)," 2020, no. August 28th (2013), https://www.dfo-mpo.gc.ca/aquaculture/sci-res/imta-amti/imta-amti-eng.htm.

28. Luana Granada et al., "Is Integrated Multitrophic Aquaculture the Solution to the Sectors' Major Challenges? – a Review," *Reviews in Aquaculture* 8, no. 3 (2016).
29. CMFRI, "Integrated Multi-Trophic Aquaculture Systems (Imta)," (2016), http://eprints.cmfri.org.in/10666/1/7.%20Geetha%20Sasikumar.pdf.
30. ASTRAL Consortium, "All Atlantic Ocean Sustainable, Profitable and Resilient Aquaculture," https://www.astral-project.eu/.
31. Cordis, "All Atlantic Ocean Sustainable, Profitable and Resilient Aquaculture," https://cordis.europa.eu/project/id/863034.
32. Kiran Dube and Thongam Chanu, "Organic Aquaculture: Way to Sustainable Production," in *Advances in Fish Research* (Delhi, India: Narendra Publishing House, 2012).
33. Xie Biao, "The Development of and Prospects for Organic Aquaculture Worldwide," *Outlook on Agriculture* 37, no. 4 (2008).
34. Asem Sanjit Singh, Manoharmayum Devi, and Atom Arun Singh, "Organic Aquaculture: Sustainable and Superior Production the Concept of Organic Aquaculture," *Journal of the World Aquaculture Society* (2011).
35. Claude Boyd and Aaron McNevin, *Aquaculture, Resource Use, and the Environment* (Somerset, UNITED STATES: John Wiley & Sons, Incorporated, 2015).
36. Ratheesh R. et al., "Organic Aqua-Farming: A Gateway to Sustainable Aquaculture," *Aquainternational* (2013).
37. European Commission, "10 Years of the Eu Organic Logo," https://ec.europa.eu/info/food-farming-fisheries/farming/organic-farming/organic-logo_en.
38. Nuno Sales Henriques et al., "Marxan as a Zoning Tool for Development and Economic Purposed Areas - Aquaculture Management Areas (Amas)," *Ocean & Coastal Management* 141 (2017).
39. FAO and World Bank, "Aquaculture Zoning, Site Selection and Area Management under the Ecosystem Approach to Aquaculture," (2015), http://www.fao.org/3/a-i5004e.pdf.
40. Sustainable Fisheries Partnership, "Best Practices for Aquaculture Management: Guidance for Implementing the Ecosystem Approach in Indonesia and Beyond," (2018), https://www.sustainablefish.

org/News/SFP-Conservation-International-and-UCSB-release-new-best-practices-for-aquaculture-management-guide.
41. Ibid.
42. Weitzman and Filgueira, "The Evolution and Application of Carrying Capacity in Aquaculture: Towards a Research Agenda."
43. Sustainable Fisheries Partnership, "Best Practices for Aquaculture Management: Guidance for Implementing the Ecosystem Approach in Indonesia and Beyond."
44. FAO and World Bank, "Aquaculture Zoning, Site Selection and Area Management under the Ecosystem Approach to Aquaculture: A Handbook," (2017), http://www.fao.org/3/i6992e/i6992e.pdf.
45. Ibid.
46. Sustainable Fisheries Partnership, "Best Practices for Aquaculture Management: Guidance for Implementing the Ecosystem Approach in Indonesia and Beyond."
47. Ibid.
48. Department of Primary Industries and Regional Development, "Aquaculture Development Zones," 2021, http://www.fish.wa.gov.au/fishing-and-aquaculture/aquaculture/aquaculture%20zones/Pages/default.aspx.
49. Cecile Brugère et al., "The Ecosystem Approach to Aquaculture 10 years on - a Critical Review and Consideration of Its Future Role in Blue Growth," *Reviews in Aquaculture* 11, no. 3 (2019).
50. FAO and World Bank, "Aquaculture Zoning, Site Selection and Area Management under the Ecosystem Approach to Aquaculture."
51. Ibid.
52. Ibid.
53. Sustainable Fisheries Partnership, "Best Practices for Aquaculture Management: Guidance for Implementing the Ecosystem Approach in Indonesia and Beyond."
54. Primary Industries and Regions SA, "The Sustainable Management of South Australian Aquaculture," (2014), https://www.pir.sa.gov.au/__data/assets/pdf_file/0007/221992/Aquaculture_SustainMgmet.pdf.

References

Amirkolaie, Abdolsamad K. "Reduction in the Environmental Impact of Waste Discharged by Fish Farms through Feed and Feeding." *Reviews in Aquaculture* 3, no. 1 (2011/03/01/ 2011): 19–26.

ASTRAL Consortium. "All Atlantic Ocean Sustainable, Profitable and Resilient Aquaculture." https://www.astral-project.eu/.

Biao, Xie. "The Development of and Prospects for Organic Aquaculture Worldwide." *Outlook on Agriculture* 37, no. 4 (2008): 255–60.

Boyd, Claude, and Aaron McNevin. *Aquaculture, Resource Use, and the Environment.* Somerset, UNITED STATES: John Wiley & Sons, Incorporated, 2015.

Brugère, Cecile, José Aguilar-Manjarrez, Malcolm C. M. Beveridge, and Doris Soto. "The Ecosystem Approach to Aquaculture 10 years on - a Critical Review and Consideration of Its Future Role in Blue Growth." *Reviews in Aquaculture* 11, no. 3 (2019): 493–514.

Buck, Bela H., Max F. Troell, Gesche Krause, Dror L. Angel, Britta Grote, and Thierry Chopin. "State of the Art and Challenges for Offshore Integrated Multi-Trophic Aquaculture (Imta)." [In English]. *Frontiers in Marine Science* 5, no. 165 (2018/05/15/ 2018).

Byron, C., and Barry Costa-Pierce. "Carrying Capacity Tools for Use in the Implementation of an Ecosystems Approach to Aquaculture." *FAO Fisheries and Aquaculture Proceedings* 21 (01/01/ 2013): 87–102.

CMFRI. "Integrated Multi-Trophic Aquaculture Systems (Imta)." (2016). http://eprints.cmfri.org.in/10666/1/7.%20Geetha%20Sasikumar.pdf.

Cordis. "All Atlantic Ocean Sustainable, Profitable and Resilient Aquaculture." https://cordis.europa.eu/project/id/863034.

Correia, Marta, Isabel Costa Azevedo, Helena Peres, Rui Magalhães, Aires Oliva-Teles, Cristina Marisa Ribeiro Almeida, and Laura Guimarães. "Integrated Multi-Trophic Aquaculture: A Laboratory and Hands-on Experimental Activity to Promote Environmental Sustainability Awareness and Value of Aquaculture Products." [In English]. *Frontiers in Marine Science* 7, no. 156 (2020/03/20/ 2020).

CTA Brussels Office. "Fish-Farming: The New Driver of the Blue Economy." (2013). https://europa.eu/capacity4dev/file/15656/download?token=J7_MU3_d.

Dauda, Akeem Babatunde, Abdullateef Ajadi, Adenike Susan Tola-Fabunmi, and Ayoola Olusegun Akinwole. "Waste Production in Aquaculture:

Sources, Components and Managements in Different Culture Systems." *Aquaculture and Fisheries* 4, no. 3 (2019/05/01/ 2019): 81–88.

Department of Primary Industries and Regional Development. "Aquaculture Development Zones." 2021, http://www.fish.wa.gov.au/fishing-and-aquaculture/aquaculture/aquaculture%20zones/Pages/default.aspx.

Dube, Kiran, and Thongam Chanu. "Organic Aquaculture: Way to Sustainable Production." In *Advances in Fish Research*, 219–229. Delhi, India: Narendra Publishing House, 2012.

European Commission. "10 Years of the Eu Organic Logo." https://ec.europa.eu/info/food-farming-fisheries/farming/organic-farming/organic-logo_en.

FAO. "Aquaculture." http://www.fao.org/aquaculture/en/.

———. "Mariculture as a Method of Using Coastal Zones." http://www.fao.org/3/AC583E/AC583E00.htm#ch2.

———. "Planning and Management for Sustainable Coastal Aquaculture." (2001). http://www.fao.org/3/a-y1818e.pdf.

———. "Rural Aquaculture: Overview and Framework for Country Reviews." (1998). http://www.fao.org/3/x6941e/x6941e04.htm#bm04.2.

———. "Sustainable Development of Marine Aquaculture Off-the-Coast and Offshore – a Review of Environmental and Ecosystem Issues and Future Needs in Temperate Zones." (2013). http://www.fao.org/tempref/FI/CDrom/P24/i3530e/root/10.pdf.

FAO and World Bank. "Aquaculture Zoning, Site Selection and Area Management under the Ecosystem Approach to Aquaculture." (2015). http://www.fao.org/3/a-i5004e.pdf.

———. "Aquaculture Zoning, Site Selection and Area Management under the Ecosystem Approach to Aquaculture: A Handbook." (2017). http://www.fao.org/3/i6992e/i6992e.pdf.

Fisheries and Oceans Canada. "Aquaculture in Canada: Integrated Multi-Trophic Aquaculture (Imta)." 2020, no. August 28th (2013). https://www.dfo-mpo.gc.ca/aquaculture/sci-res/imta-amti/imta-amti-eng.htm.

Free, Christopher M., Tracey Mangin, Jorge García Molinos, Elena Ojea, Merrick Burden, Christopher Costello, and Steven D. Gaines. "Realistic Fisheries Management Reforms Could Mitigate the Impacts of Climate Change in Most Countries." *PLOS ONE* 15, no. 3 (2020): e0224347.

Granada, Luana, Nádia Sousa, Sofia Lopes, and Marco F. L. Lemos. "Is Integrated Multitrophic Aquaculture the Solution to the Sectors' Major Challenges? – a Review." *Reviews in Aquaculture* 8, no. 3 (2016/09/01/ 2016): 283–300.

Henriques, Nuno Sales, Pedro Monteiro, Luis Bentes, Frederico Oliveira, Carlos M. L. Afonso, and Jorge M. S. Gonçalves. "Marxan as a Zoning Tool for Development and Economic Purposed Areas - Aquaculture Management Areas (Amas)." *Ocean & Coastal Management* 141 (2017/06/01/ 2017): 90–97.

Meta. "Future Farming: Fishing for a Sustainable Seafood Model in Spain." https://meta.eeb.org/2020/07/23/future-farming-fishing-for-a-sustainable-seafood-model-in-spain/.

National Science Challenges. "Let's Turn Those Buoys into Real-Time Mussel Sensors." https://www.sftichallenge.govt.nz/news/lets-turn-those-buoys-real-time-mussel-sensors/.

Primary Industries and Regions SA. "The Sustainable Management of South Australian Aquaculture." (2014). https://www.pir.sa.gov.au/__data/assets/pdf_file/0007/221992/Aquaculture_SustainMgmet.pdf.

R. Ratheesh, Sandeep K. P., Manju Lekshmi, and Sreekanth G. B. "Organic Aqua-Farming: A Gateway to Sustainable Aquaculture." *Aquainternational* (08/01/ 2013): 25–28.

Rosa, João, Marco F. L. Lemos, Daniel Crespo, Margarida Nunes, Andreia Freitas, Fernando Ramos, Miguel Ângelo Pardal, and Sara Leston. "Integrated Multitrophic Aquaculture Systems – Potential Risks for Food Safety." *Trends in Food Science & Technology* 96 (2020/02/01/ 2020): 79–90.

Sanjit Singh, Asem, Manoharmayum Devi, and Atom Arun Singh. "Organic Aquaculture: Sustainable and Superior Production the Concept of Organic Aquaculture." *Journal of the World Aquaculture Society* (09/01/ 2011).

Sustainable Fisheries Partnership. "Best Practices for Aquaculture Management: Guidance for Implementing the Ecosystem Approach in Indonesia and Beyond." (2018). https://www.sustainablefish.org/News/SFP-Conservation-International-and-UCSB-release-new-best-practices-for-aquaculture-management-guide.

Tsani, Stella, and Phoebe Koundouri. "A Methodological Note for the Development of Integrated Aquaculture Production Models." [In English]. *Frontiers in Marine Science* 4, no. 406 (2018/01/04/ 2018).

Veta la Palma. "Extensive Aquaculture." https://www.vetalapalma.es/en/extensive-aquaculture/.

Waikato Regional Council. "Monitoring Framework for the Waikato Coastal Marine Area: Report 3 – Seabed and Water Column Monitoring and Standards." (2015). https://www.waikatoregion.govt.nz/services/publications/tr201540/.

Weitzman, Jenny, and Ramón Filgueira. "The Evolution and Application of Carrying Capacity in Aquaculture: Towards a Research Agenda." *Reviews in Aquaculture* (2019).

6

Marine Biotechnology

Introduction

Biotechnology is a technology that utilises biological systems, living organisms or parts of this to develop or create different products. In the context of the blue economy, biotechnology advances can increase food security (through, for example, new vaccines), improve human health (such as biomedical, pharmaceutical and nutraceutical applications with bioactive compounds), foster environmental recovery and restoration (through, for example, marine organism-based bioremediation), produce renewable energy (such as biogas). Furthermore, a range of non-energy products can be derived from macroalgae.[1] This chapter will first introduce the concept of marine biotechnology before discussing its application in food security, human health, and environmental recovery and restoration. The chapter will then discuss the use of microalgae and macroalgae in biofuel production. Finally, the chapter will discuss the range of supplementary products derived from macroalgae.

R. C. Brears, *Developing the Blue Economy*,
https://doi.org/10.1007/978-3-030-84216-1_6

Marine Biotechnology

Marine biotechnology can be viewed as the use of marine bioresources as the target or source of biotechnological applications and includes:

- Marine organisms used as feedstock, for example, to produce food, fuel, materials, or bioactive compounds
- Products extracted from marine organisms or developed in laboratories using the knowledge of the natural processes or properties of marine organisms, including products created from marine DNA via genetic engineering or synthetic biology
- Processes catalysed by marine organisms or derivatives thereof
- Ecosystem services such as biosensors and bioremediation
- The application of biotechnology knowledge to fish health and welfare
- Understanding and mapping of ecosystems based on generic biotechnological tools and knowledge[2]

Food Security

Aquaculture plays a vital role in food security and livelihood and is a source of income and social development in many countries. Nonetheless, loss of production in aquaculture occurs, with disease being the most serious constraint that results in food insecurity, loss of farmers' livelihoods, loss of jobs, and reduced incomes.[3,4] Fish immunisation has been carried out for over 50 years and is generally an accepted method for preventing a wide range of bacterial and viral diseases. Vaccination efforts contribute to environmental, social, and economic sustainability in global aquaculture. In many parts of the world, there has been a transition away from antibiotics and towards vaccination. A typical fish vaccination either contains or produces a substance that serves as an antigen. This component then stimulates an innate and/or adaptive immune response within the fish against a particular pathogen.

Conventional Fish Vaccines

Conventional fish vaccines primarily consist of inactivated whole organisms and live vaccines. The result has been increased production for commercial aquaculture and reduced chemical therapeutics and feed delivered antibiotics.

Inactivated Vaccines

Inactivated vaccines are created from a virulent disease-causing microbe that has been through a process to lose its ability to infect or replicate in or outside a host. The changes can be induced through physical, chemical, or radiation processes without compromising the antigenicity of the microbial agent. Inactivated vaccines are more stable than live vaccines and can be less expensive to produce. Inactivated vaccines do not persist in the environment and may induce weaker or shorter-lived immunity when compared to other vaccine types.

Live Vaccines

Modified live vaccines are prepared from one or more viruses or bacteria displaying attenuated virulence or natural low virulence toward the target fish species. Pathogens can be attenuated using physical or chemical processes, serial passage in cell culture, culture under abnormal conditions, or genetic manipulation. Live vaccines are usually more immunogenic than killed preparations due to their ability to proliferate or enter the host and stimulate greater cellular responses linked to both innate and adaptive immunity.

Alternative Vaccine Technology

Alternative vaccine technologies have been commercialised too, including:

- *Subunit vaccines*: A subunit vaccine is a fragment of a pathogen and is typically a surface protein used to trigger an immune response and stimulate acquired immunity against the pathogen from which it is derived. Subunit vaccines cannot replicate in the host. There is no risk of pathogenicity to the host or non-target species
- *Nucleic acid vaccines*: Nucleic acid vaccines consist of DNA or RNA and are simple to generate and safe to administer since they cannot revert to a pathogenic state
  - DNA vaccines involve the direct introduction into appropriate tissues of a plasmid containing the DNA sequence encoding the antigen(s) against which an immune response is sought and relies on in situ production of the target antigen
  - RNA vaccines work by introducing a messenger RNA (mRNA) sequence coded for a disease-specific antigen. Once the mRNA strand in the vaccine is inside the body's cells, the cells use the genetic information to produce the antigen. This antigen is displayed on the cell surface, which is recognised by the immune system[5,6]

Vaccine Delivery Methods/Routes of Administration

Vaccines can be administered to fish through three different routes as follows.

Oral Vaccination

Antigen can be introduced to fish by direct delivery via the fish's digestive system at any age. It is the easiest method logistically because feeding is a routine practice in fish farms. In oral vaccination, the vaccine is either mixed with the feed, top dressed on the feed, sprayed over the feed, or bioencapsulated. Delivery of antigen in fish feed offers various advantages, including cost-effectiveness, ease of safe administration in all sizes/stages of fish, and imposing low stress. Nonetheless, oral application of vaccines induces low protection levels and a relatively short duration of protection due to degradation of the antigens in the gastrointestinal tract, for instance. Oral vaccinations can be administered for primary vaccination or as a booster vaccine to develop protection against long-lasting endemic diseases.[7]

Injection Vaccination

Only a small, identified concentration of antigen can be directly injected into fish by intraperitoneal (IP) and intramuscular routes, with the most efficient method of fish immunisation being IP. Injection vaccinations are performed manually using a needle or device such as compressed air. The antigens can easily be stored at four degrees Celsius for injectable vaccines. The method also allows for the simultaneous delivery of multiple antigens from different pathogens in the form of a multivalent vaccine. However, this method is not suitable for fish weighing less than five kilograms. It is very labour intensive, results in the temporary reduction in feeding, can accidentally puncture the intestine, and can cause a wound at the injection site, resulting in a secondary infection.[8]

Immersion Vaccination

Immersion vaccines work on the ability of mucosal surfaces to recognise pathogens that they have been in contact with. When fish are immersed in water containing the diluted vaccine, the suspended antigens from the vaccine may be absorbed by the skin and gills. Then, specialised cells, such as antibody-secreting cells present in the skin and gill epithelium, will be activated and protect the fish when exposed to the live pathogen at a later stage. Immersion vaccination (short or long bath) is recommended for smaller fish weighing between one and four grams. The method is rapid, effective, less stressful, convenient, and economical to vaccinate fish that require minimal handling stress. The duration of the project can range from three to 12 months and therefore requires a booster vaccination.[9]

Case: Developing an Oral Vaccination for Salmon

Researchers from the University of Aberdeen, University of Stirling, and Queens University Belfast, with industrial partners, are researching the development of an oral vaccination for salmon. The research will include testing a novel oral vaccine delivery technology based on nanoparticles. The goal of the project is to focus on crucial areas required to understand oral vaccine efficacy, including:

- Determining how foreign molecules are recognised and presented to immune cells in the gut of the salmon
- Evaluating gene markers of vaccine effectiveness using existing commercial vaccines that use mucosal delivery, either by immersion of fish in the vaccine solution as a primary vaccination or given as oral boosters
- Evaluating the use of silicon-based nanoparticles to deliver vaccines against two commercially relevant diseases[10]

Human Health

One of the twentieth century's significant accomplishments was developing modern pharmaceuticals, with new drug therapies extending the human life span and improving the quality of life. Marine plants, animals, and microbes produce compounds that have the potential as pharmaceuticals. Over 30,000 compounds of marine origin are known, and since 2008, over 1,000 compounds are newly discovered each year. The compounds are typically characterised by structural novelty, complexity, and diversity.[11] The compounds mainly sought are ones with cytostatic and cytotoxic activity, with anti-cancer compounds representing more than half of new marine natural products discovered from 1985 to 2012, followed by compounds with anti-inflammatory, antifungal, antiviral, and analgetic properties.[12]

Marine Drug Development

The marine drug development process starts with collecting and identifying marine organisms, with the collection of organisms at locations that are easily reached preferred, such as coastlines or from the water surface. If possible, collected and identified organisms should be taken into cultivation. If not, enough purified and dried material from the collection is necessary for further analysis. The following steps are the same as for terrestrial organisms: (1) extraction; biological and/or chemical and/or genetic screening; (2) identification and selection of interesting extracts; (3) isolation and structure elucidation of compounds with promising biological activity and/or novel structures; (4) broad pharmacological and toxicological investigations of drug candidates for pharmacodynamics (the study of the biochemical and physiologic effects of drugs), pharmacokinetic (refers to the movement of the drug into, through, and out of the body) and safety parameter; (5) target identification; (6) ensuring of supply; (7) possibly derivatisation and production of substance libraries and selection of best suitable candidate; (8) clinical trials; and (9) registration according to the valid legislation.[13]

Supply

A critical aspect of drug development from marine organisms is the permanent availability of sufficient amounts of organisms and compounds without harming the marine environment. If collection from the natural environment cannot be done sustainably, the supply problem can be solved by processes of marine biotechnology:

- *Aquaculture/cultivation*: Most pharmaceutically interesting marine organisms cannot be cultured under artificial conditions. As such, a better understanding of living conditions in the natural environment is necessary to develop alternative cultivation methods and to maintain metabolite production over a long time
- *Genetic engineering*: This involves the transfer of genetic information from the desired compound into host cells, which can be more easily

cultivated, and the sustainable production of the compound in the host cells. Currently, this approach is realised at the research level but not at the industrial scale for marketable marine drugs

- *Synthesis or semi-synthesis*: While total synthesis is principally possible for many known marine compounds, it can only be economically realised for relatively simple products. Another way can be semisynthetic production, where easily available compounds are transferred by chemical or biochemical processes into the desired product[14]

Nutraceuticals

The concept of nutraceutical comes from the combining of "nutrition" and "pharmaceutical". The term refers to raw foods, fortified foods, or dietary supplements containing biologically active molecules, also known as bioactive molecules, that provide health benefits beyond basic nutrition, including antioxidant, antibiotic, antiparasitic, antiviral, anti-inflammatory, antifibrotic, and anti-cancer benefits. These bioactive molecules can be obtained either by extraction from natural sources or by chemical and biotechnological synthesis. Marine organisms are a valuable source of bioactive molecules that provide an unlimited resource for developing new bioactive products, summarised in Table 6.1.[15,16,17,18]

Case: Bioactive Seaweed-based Ingredients and Products in the Nordic Region

Nordic Innovation ran a project to develop technologies to process bioactive ingredients from bladderwrack and develop innovative products containing them. The project's objective was to create new high-value ingredients and products from marine seaweeds, currently an underutilised resource in the Nordic region. Furthermore, the project's objective was to start the commercial production of novel bioactive marine seaweed-based ingredients and products containing them, including food supplements, cosmetics, and food antioxidants. The project developed extraction methods and analysed and tested the composition of the extracts and their stability, resulting in know-how in manufacturing and

Table 6.1 Marine Organisms Providing a Resource for Bioactive Products

Marine organism	Description
Marine algae	• Marine algae are used as sources of food and food ingredients • Microalgae are rich sources of food ingredients, such as Vitamins C, A, H, B1, B2, B6, and B12, and therefore bioactive molecules from microalgae are commercially produced, used as food additives, and incorporated into infant milk formulations and dietary supplements • Macroalgae, also called seaweed, are the most popular type of algae as it provides a large variety of food and food ingredients. Macroalgae are important sources of essential nutrients and new bioactive molecules for human nutrition. For example, red and brown seaweeds are alternative sources of vitamins, minerals, and proteins and are good sources of essential fatty acids
Marine fishes	• Fish consumption provides a range of nutritional benefits due to the presence of proteins, unsaturated essential fatty acids, minerals (for example, calcium, iron, selenium, and zinc), and vitamins, particularly Vitamin A, B3, B5, B12, E, and D
Marine invertebrates	• Marine invertebrates, such as sponges, molluscs, echinoderms, and crustaceans, are sources of bioactive peptides and steroids. Extracts or compounds isolated from marine invertebrates are known to have antibacterial, antiviral, anthelmintic, antifungal properties, for example
Sponges	• Sponges contain many chemically diverse compounds like alkaloids, terpenoids, polyketides, macrolides, polyphenolic compounds, peptides, and sterols that are isolated from sponges with the potential to cure various ailments

(continued)

Table 6.1 (continued)

Marine organism	Description
Molluscs, Echinoderms, and Crustaceans	• Molluscs, together with echinoderms, have been widely consumed as marine foods and are considered natural functional foods. Bioactive peptides obtained from the fermented blue mussel and oyster sauces significantly decrease hypertension, while ground abalone and its shells are used for treating eye diseases • Nutrient composition of marine crustaceans like shrimp and krill decrease the total blood lipids in humans and improve Vitamin A levels, specific proteins, and eicosapentaenoic acid, an omega-3 fatty acid

the use of seaweeds. Based on this work, a marketing and sales plan was created for a line of high-end marine bioactive skincare products named UNA Skincare and MarinoxTM.[19]

Cosmetics

The European Commission defines cosmetic products as "*any substance or mixture intended to be placed in contact with the external parts of the human body (epidermis, hair system, nails, lips and external genital organs) or with the teeth and the mucous membranes of the oral cavity with a view exclusively or mainly to cleaning them, perfuming them, changing their appearance, protecting them, keeping them in good condition or correcting body odours*".[20] Marine resources are becoming an increasingly significant source of active ingredients for the cosmetics industry.

Moisturising Care

Maintenance of the hydration rate is essential to preserve skin integrity. Marine organisms produce several molecules with moisturising properties such as polysaccharides, fatty acids (sophorolipids, rhamnolipids, and mannosylerythritol), and proteins that are widely used on the skin. For instance, marine fish proteins mainly consist of collagen, which is widely used in cosmetics for its moisturising properties. Skin-hydrating and skin-firming cosmetics formulated with fish-derived collagen have been evaluated. Low doses of collagen hydrolysates derived from jellyfish have also demonstrated their potential for moisturising agents.[21,22]

Active Ingredients to Prevent Skin Ageing

While genetic factors play a role in skin ageing, so do environmental factors, particularly UV exposure and weather (wind exposure, etc.). Among the bioactive substances with anti-ageing action of marine origin, bacterial polysaccharides (PSs) are most used. PSs are produced by microalgae. Marine-derived collagen is widely used in cosmetic formulations due to its repair and regeneration properties. Seawater contains minerals, such as sodium, potassium, magnesium, calcium, sulphates, and chlorides, which are beneficial for the skin.[23,24]

Active Ingredients for Topical Photoprotection

The skin has three layers that act as a chemical and physical barrier. The skin can be damaged by various environmental factors, including chemicals, UV, and pollution. Prolonged exposure of skin to UV radiation can lead to photo-induced skin ageing and photo-carcinogenesis. Several marine organisms produce UV-absorbing compounds to protect themselves from UV radiation. Therefore, marine organisms are a significant source of photo-protective compounds in cosmetics, including sunscreen. Mycosporine-like amino acids (MAAs) are intracellular water-soluble colourless compounds found in many marine and freshwater organisms. MAAs absorb UV radiation and dissipate this energy in the form of heat radiation to the surrounding environment.[25,26]

Active Ingredients with Skin Whitening Properties

Most skin-whitening compounds used in cosmetics are still provided by terrestrial organisms, providing an opportunity for marine skin whitening molecule-based research in cosmetics. For instance, *Pistacia lentiscus*, found in saline environments, contains gallic acids and epicatechins, which could effectively treat hyperpigmentation.[27,28]

Case: Pôle Mer Bretagne Atlantique Sea Innovation Cluster

The Pôle Mer Bretagne Atlantique is a sea innovation cluster in France containing more than 390 cluster members, 220 small and mid-size enterprises, 45 large companies, 52 research organisations, universities, technology centres, and 41 other ecosystem actors. The cluster is focused on six strategic areas, including marine biological resources. Three themes underpin innovation in this area: sustainable fishing, sustainable aquaculture, and marine biotechnology. Blue biotech encompasses the entire value chain from understanding and characterising the potential of marine biodiversity to produce effective, natural, active ingredients to developing biotechnological industrial processes that provide clean and profitable extraction and production of molecules/extracts using marine biomass. A variety of blue biotech projects being, or have been, undertaken include:

- The ALGOLIFE project is developing the commercialisation of bioactive molecules extracted from macroalgae for nutrition and animal health/nutrition markets
- The AZOSTIMER project was to provide an innovative response to the problems posed by nitrogen-based fertilisers by developing new, natural fertilisers based on algae
- The DIESALG project sought to develop high-energy value microalgae to produce biodiesel
- The ODONTOMER project set out to use seawater and seaweed extracts as the ingredients for oral and dental hygiene products, such as toothpaste and mouthwash
- The PHARMASEA project was to characterise several families of molecules of interest in the study and treatment of Alzheimer's disease[29]

Environmental Recovery and Restoration

Marine biotechnology is playing an increasingly important role in the protection and management of the marine environment. Examples of these include the following.

Marine-derived Antifouling Strategies

In the marine environment, all hard surfaces, including macroorganisms, are colonised by microorganisms, mainly from the surrounding environment. Biofouling (accumulation of organisms) involves an initial biofilm formation (consisting of microbes and microalgae), followed by the settlement of invertebrate larvae and algal spores. Biofouling assemblages in the marine environment are made up of marine organisms such as bacteria, fungi, phytoplankton, polychaetes, barnacles, molluscs, ascidians, and algae. Biofouling is an ongoing problem in marine sectors as it requires controlling and cleaning processes. The effects of biofouling include loss of productivity of aquaculture or increased fuel costs to shipping and associated ongoing prevention, management, and control costs. In the past, tributyltin-containing antifouling paints were widely used in commercial vessels to control biofouling. However, due to its toxicity, it was banned internationally in 2008. As such, research has increased over the years on effective eco-friendly antifoulants for marine applications.[30] Ecological antifoulants mimic the defence mechanisms found in nature, including developing coatings that imitate natural antifouling surfaces, such as the skin of a shark. Other approaches combine natural products. For example, the molecules extracted from plants, marine organisms, bacteria, or fungi interfere with the chemical cues of the foulants. In general, these natural solutions inhibit either the attachment or growth of the organisms, preventing bioaccumulation. A couple of examples of approaches that are inspired by or mimic natural surfaces and materials include:

- *Surface topography as a strategy for antifouling*: The typography of a surface dictates its roughness and wettability. These features affect bioadhesion by either inhibiting or promoting it. Many plants and marine organisms have micro/nanotopographies that protect them from colonising organisms. As such, scientists can mimic these topographies to create surfaces with antifouling activity
- *DOPA-based antifouling materials*: There has been research into designing coatings that use bioadhesives, such as l-3,4-dihydroxyphenylalanine (L-DOPA). L-DOPA is the main component of the adhesive proteins of marine mussels. These proteins can adhere to almost any substrate (glass, metals, and Teflon) and survive harsh conditions, such as tide and high salt concentrations. Utilising these

properties, antifouling compounds can be created that adhere to a variety of surfaces[31]

Bioremediation of Marine Ecosystems

Petroleum hydrocarbons in crude oil are natural products derived from aquatic algae laid down between 180 and 85 million years ago. Crude oils regularly escape into the environment from underground reservoirs. Because petroleum hydrocarbons occur naturally in all marine environments, there is a diverse range of microorganisms that have evolved to utilise hydrocarbons as sources of carbon and energy for growth. As such, many species of bacteria, archaea, and fungi can degrade petroleum.[32,33] Bioremediation uses natural or genetically manipulated microorganisms to treat oil spills. It comprises two distinct operations:

- *Bioaugmentation*: This is the inoculation of exogeneous microorganisms into the polluted site. The success of bioaugmentation depends on the adaptation of the microorganisms to the site that needs decontamination. The success of the process also relies on the success of the introduced microorganisms to compete with the indigenous microorganisms, predators, and various abiotic factors
- *Biostimulation*: This relies on the already existing (native) microorganisms, which may be enhanced in their activities through fertilisation with N- and P- compounds. Typically, they are applied in the field by spraying aqueous nutrient solutions or spreading dry granules to enhance oil biodegradation. Nonetheless, it is difficult to achieve in non-sheltered marine environments or medium to high-energy shorelines[34,35]

Biosensing

As the ocean is vital in terms of food supply, raw material extraction, waste product disposal, transport, and recreational use, scientific researchers, as well as legislative requirements, drive the need to investigate marine processes and monitor inputs and temporal trends as well as the subsequent fate and impact of releases. For instance, there is the need for discharge monitoring of coastal industrial sites and offshore installations and the monitoring of toxic algae and their subsequent accumulation in filter feeders to protect human health. Biosensors are devices that can be used to detect the presence or concentration of a biological analyte, such as a biomolecule, biological structure, or microorganism. Biosensors consist of three parts: a component that recognises the analyte and produces a signal, a signal transducer, and a reader device. Biosensors can be used for various marine measurements, including eutrophication, organism detection, pollution, trace metals, and ecotoxicity.[36,37]

Case: Marine Environmental In situ Assessment and Monitoring Tool Box

A research consortium consisting of the Italian National Research Council, Spanish Council for Scientific Research, Norwegian Institute for Water Research, and the University of Naples developed a marine environmental in situ assessment and monitoring toolbox (MariaBox). MariaBox is a wireless marine environmental analysis device for monitoring chemical and biological pollutants when installed into a buoy. The device includes a sensing and analysis box, a modular communication system, a flexible power system, a software platform, and a cell phone application. The box transmits collected data in real-time through different channels according to local needs and geographical location, such as radio, GSM/GPRS/3G, Wi-Fi, WiMax, or satellite link. The unit is designed to be remotely controlled and allows the user to update the device's firmware and modify various configuration parameters wirelessly, significantly reducing maintenance costs. Biosensors have been developed for five human-made chemicals and four categories of microalgae toxins relevant to shellfish and fish farming.[38]

Microalgae and Macroalgae Biofuel Production

Marine algae include microalgae and cyanobacteria. Microalgae comprise unicellular plants that can be grown under natural or artificial light. Microalgae can be grown at sea in semi-porous containers nearshore, saving space, utilising natural sunlight for growth, and reducing the need for supplemental artificial nutrients. Macroalgae, commonly known as seaweeds, are typically cultivated offshore or near coastal facilities. Seabed cultivation involves seaweed pieces anchored to sandy or muddy bottoms of shallow lagoons and bays and harvested several months after planting. The crop is typically wholly or partially collected, with partial collection providing material for the next cultivation cycle. Seaweeds can also be grown on the seabed within fences without being fixed to the bottom, while line/rope cultivation involves seaweed being fixed on ropes suspended at the surface of the water or several metres below.

Microalgae and macroalgae can be grown for biofuel production, including biogas produced by anaerobic digestion of biomass, biodiesel produced from lipids accumulated in cells of algae, ethanol, hydrogen from photobiological transformations, or algae biomass that is used for direct combustion. Table 6.2 lists the main differences between microalgae and macroalgae in biofuel production.[39,40]

Microalgae for Biofuel Production

Microalgae-based biofuels are eco-friendly and non-toxic. Most microalgae species are favourable for biodiesel production due to high lipid contents (50–70% and may reach 80%). Microalgae can produce algal oil of 58,700 L/hac (L/hac), producing over 121,000 L/hac of biodiesels. Also, microalgae possess high contents of different carbohydrates, such as glycogen, starch, agar, and cellulose, which can be easily converted to fermentable sugars for bioethanol production.[41] There are different ways of converting microalgae biomass into biofuels and energy:

Table 6.2 Differences Between Microalgae and Macroalgae in Biofuel Production

Parameter	Microalgae	Macroalgae
Size	Microscopic plant-like organisms s (± 1 to 50 μm), which can be seen under a microscope	Macroalgae can reach sizes of 60 m in length
Physical structure	Microalgae do not have roots, stems, and leaves	Macroalgae are composed of a thallus and sometimes a stem and a foot
Energy density	High	Medium
Biomass yield	High	Medium
Ease of cultivation	Cultured in photobioreactors or open ponds	Cultured in natural environments such as the ocean
Ease of harvesting	Difficult	Easy
Oil yield	Produce comparatively high amounts of lipids	Produce only small amounts of lipids

- *Anaerobic digestion*: This process involves transforming microalgae into biogas by bacterial fermentation in the absence of air. This biogas is a conventional fuel that can be stored and distributed
- *Hydrothermal liquefaction*: This consists of transforming biomass in a complex process involving chemical and physical structural changes. Biomass is broken down into small molecules. These small molecules are unstable and reactive and can be repolymerised into oily compounds with a wide distribution of molecular weights. It does not require drying of the raw material (energy expenditure of water evaporation is avoided). The process occurs between 250–350 degrees Celsius. Generally, bio-oil is obtained after the extraction and evaporation of an organic solvent
- *Pyrolysis*: This is the chemical decomposition of a substance under the action of heat. In the context of biomass, it is converted into bio-oil, syngas, and a carbon residue, at high temperatures (350–700 degrees Celsius), in the absence of air and under low pressure. Short residence times, rapid heating rates, and moderate temperatures promote liquid product yield, with slow pyrolysis promoting the formation of

carbonaceous residues and fast pyrolysis promoting the formation of bio-oil
- *Gasification*: This is a process that converts carbonaceous materials, such as coal, oil, or biomass, into carbon monoxide and hydrogen by reaction of the raw material with a controlled amount of oxygen at extremely high temperatures (800–1,000 degrees Celsius)[42]

Some of the benefits of using microalgae for biofuel production include the following:

- The harvesting and transporting of microalgae are relatively low compared to the costs of using terrestrial plant biomass resources. Also, the production cycle does not affect the human food chain supply system, reducing food-energy pressures
- Microalgae do not compete with land-based plants used for food production, fodder, and other value-add products
- Microalgae can grow in fresh, brackish, or saltwater environments or non-arable lands that are not compatible for growing other crops and conventional agriculture
- Microalgae contain oil that can range from 20–50% of the dry weight of biomass
- Microalgae can produce various valuable supplementary products, including carbohydrates, proteins, biopolymers, and residual biomass, and these can be used for feed or fertiliser purposes. Furthermore, herbicides and pesticides are not required in the cultivation of microalgae
- Microalgae can fix carbon dioxide in the atmosphere. Microalgae can also affect the biofixation of waste carbon dioxide[43]

Macroalgae for Biofuel Production

The primary way of categorising the extraction of energy from macroalgae is whether an initial drying step is required or not. This leads to two types of processes:

1. *Energy extraction methods requiring dry macroalgae*: The energy extraction methods requiring dry macroalgae include:
    a. *Direct combustion*: This is the primary method by which energy from biomass resources is realised, providing heat or steam for household and industrial uses or the production of electricity
    b. *Pyrolysis*: This is the thermal decomposition of organic compounds of dry biomass by heating in the absence of air. Pyrolysis processes can be classified by temperature and process time:
        i. *Slow*: This has long residence times (from minutes to days for solids) at low reactor temperatures (<400 degrees Celsius) with extremely low rates of heating (0.01–2 °C·s − 1) and results in higher yields of char rather than the liquid or gaseous fuel products
        ii. *Fast and flash*: Fast pyrolysis is characterised by moderate pyrolysis treatment temperatures (400–600 degrees Celsius), rapid heating rates of the biomass particles (>100 degrees Celsius/minute), combined with short residence times of the biomass particles and pyrolysis vapours (0.5–2 s). Flash pyrolysis is characterised by rapid heating rates (>1,000 degrees Celsius/second), high reaction temperatures (900–1,300 degrees Celsius), and very short residence times of less than 0.5 s and can achieve greater liquid product and gas yields of around 70–80%, compared to 15–65% from slow pyrolysis. The bio-oil product generated from both these processes is more attractive than char or syngas as it has a high-energy density and is easily transported and stored
    c. *Gasification*: Gasification is the conversion of organic matter by partial oxidation at high temperature (800–1,000 degrees Celsius), mainly into a combustible gas mixture (syngas). The syngas has a mixture of hydrogen (30–40%), carbon monoxide (20–30%), methane (10–15%), ethylene (one percent), nitrogen, carbon dioxide, and water vapour. The gas can be burnt to produce heat or converted to electricity and heat in a combined gas turbine system. The syngas produced can also be used to produce methanol and hydrogen as a fuel for transport and other uses. However, the cost

of methanol produced from methane from biomass is significantly higher than from fossil fuel gas

d. *Transesterification to biodiesel*: Algal oil from macroalgae can be used to produce biodiesel through transesterification reaction. The transesterification of vegetable oils, animal fats, or waste cooking oils is the process behind conventional biodiesel. In the transesterification process, a glyceride reacts with an alcohol (typically methanol or ethanol) in the presence of a catalyst forming fatty acid alkyl esters and an alcohol

2. *Energy extraction methods for wet macroalgae*: Although drying algae can extend storage time and decrease feedstock transport cost, it requires high-energy inputs, which can be problematic if the energy source is expensive or non-renewable. There are a variety of methods that do not require drying and can utilise wet algae:

   a. *Hydrothermal treatments*: Hydrothermal subcritical water technologies utilise liquid state high-pressure–high-temperature (100 to 374 degrees Celsius) water to process biomass into a variety of products:
      i. *Hydrothermal hydrolysis*: This occurs in water heated to 100 to 240 degrees Celsius and involves the breakdown of polymers into monomers such as simple sugars, which can be fermented into organic chemicals such as ethanol, butanol, and acetone
      ii. *Hydrothermal carbonisation*: This occurs in water heated to 180 to 250 degrees Celsius and involves the carbon fraction in the solid residue (hydrochar) being enhanced, providing carbon-based products
      iii. *Hydrothermal liquefication*: This occurs at water temperatures above 280 degrees Celsius with biomass liquified to a high-energy liquid bio-oil
   b. *Bioethanol*: The process of ethanol production is commonly carried out in three steps, first, the gathering of materials containing fermentable sugars, second, the conversion of sugars into ethanol by fermentation, and third, ethanol separation and purification. The fermentation process can use any sugar-containing material to produce ethanol. Brown, green, and red

algae can be fermented to ethanol, but brown algae are suggested as the principal feedstock for bioethanol production because they have high carbohydrate contents and can be readily mass-cultivated

c. *Anaerobic digestion*: Macroalgae are potential sources of anaerobic digestion. The anaerobic digestion product is biogas, a mixture containing about 60–70% methane, 30–40% carbon dioxide, and variable trace amounts of CO, N2, O2, H2, and H2S (H2S must be removed before downstream conversion). Biogas can be exploited directly as fuel or used as raw material for the production of synthetic gas or hydrogen[44,45,46]

Some of the benefits of using macroalgae for biofuel production include the following:

- Macroalgae is highly productive, with more than five harvests possible per year
- Macroalgae can succeed in salty water with only sunlight and available nutrients from the seawater. They do not require any chemical fertiliser, saving money and energy resources
- Production of bioethanol from terrestrial plants has a significant impact on the environment from eutrophication, acidification, and ecotoxicity, all of which is mainly caused by agricultural practices
- Macroalgae can live in a variety of environmental conditions, such as along coasts
- Converting the macroalgae biomass to ethanol instead of using terrestrial plant biomass has no negative impact on food security
- Algae biomass can be cultivated in unused coastal areas, reducing land and water requirements in relation to using terrestrial plant biomass
- Several algae species can remove heavy metals from water
- In addition to bioethanol production, algae biomass can produce a wide variety of supplementary products, for instance, protein, pigments, plastics, etc.[47,48]

Case: Tugboat Trialling Algae-based Biodiesel in Japan
Mitsui OSK. Lines (MOL) and euglena Co. Ltd. conducted a successful sea trial in the Port of Nagoya, Japan of a MOL-operated tugboat using next-generation renewable biodiesel fuel developed by euglena Co. Ltd. The test, conducted with the cooperation of the Nagoya Port Authority, involved the use of renewable fuel produced with used cooking oil and *Euglena*, a type of algae-like brown seaweed and sea tangle. Marine diesel engines can run on this fuel with no modifications. Overall, the biofuel contains no sulphur and produces significantly lower greenhouse gas emissions during combustion than fossil-fuel derived fuels.[49]

Supplementary Products from Macroalgae

There are various non-energy products obtained from macroalgae, for instance:

- *Food*: The use of seaweed as a food has been traced back to the fourth century in Japan and the sixth century in China. A variety of seaweed species have been used for human consumption either traditionally or more recently, as listed in Table 6.3
- *Fertiliser*: The high fibre content of seaweed can be used as a soil conditioner and assist in moisture retention. Seaweed fertilisers can be produced in concentration form for dilution by the user. These fertilisers can be applied directly onto plants, or they can be watered in and around the root areas
- *Hydrocolloids*: Various red and brown seaweeds produce three hydrocolloids: agar, alginate, and carrageenan. A hydrocolloid is a non-crystalline substance with very large molecules that dissolve in water to give a thickened (viscous) solution. They can be used to thicken aqueous solutions, form gels (jellies) of varying degrees of firmness, form water-soluble films, and stabilise some products, such as ice cream[50,51,52,53,54]

Table 6.3 Seaweed Species for Human Consumption

Macroalgae	Food Type
Palmaria palmate	It can be eaten raw, dried or in powder form
Laminaria sp.	It can be eaten either fresh, dried, or pickled
Ascophyllum nodosum	Used as a health food
Fucus vesiculosus	Boiled and used as a health drink
Ulva lactuca	Added to soups or used in salads
Porphyra	The laver is boiled then minced to produce laverbread

Notes

1. Esra Gov and Kazim Y. Arga, "Systems Biology Solutions to Challenges in Marine Biotechnology," *Frontiers in Marine Science* 1, no. 14 (2014).
2. OECD, "Marine Biotechnology: Enabling Solutions for Ocean Productivity and Sustainability," (2013), https://www.oecd.org/greengrowth/marine-biotechnology-ocean-productivity-sustainability.htm.
3. Ayalew Assefa and Fufa Abunna, "Maintenance of Fish Health in Aquaculture: Review of Epidemiological Approaches for Prevention and Control of Infectious Disease of Fish," *Veterinary Medicine International* 2018 (2018).
4. Aslizah Mohd-Aris et al., "Live Vaccines against Bacterial Fish Diseases: A Review," *Veterinary world* 12, no. 11 (2019).
5. Jie Ma et al., "A Review of Fish Vaccine Development Strategies: Conventional Methods and Modern Biotechnological Approaches," *Microorganisms* 7, no. 11 (2019).
6. PHG Foundation, "Rna Vaccines: An Introduction," (2018), https://www.phgfoundation.org/briefing/rna-vaccines.
7. Maryam Dadar et al., "Advances in Aquaculture Vaccines against Fish Pathogens: Global Status and Current Trends," *Reviews in Fisheries Science & Aquaculture* 25, no. 3 (2017).
8. Ibid.
9. Ibid.

10. University of Aberdeen, "Development of Novel Oral Vaccination Strategies for Atlantic Salmon," https://www.abdn.ac.uk/sfirc/research/vaccine-development/noval-oral-vaccination-strategies/.
11. N.R. Council et al., *From Monsoons to Microbes: Understanding the Ocean's Role in Human Health* (National Academies Press, 1999).
12. Ulrike Lindequist, "Marine-Derived Pharmaceuticals - Challenges and Opportunities," *Biomolecules & therapeutics* 24, no. 6 (2016).
13. Ibid.
14. Ibid.
15. Hafiz A. Suleria et al., "Marine-Based Nutraceuticals: An Innovative Trend in the Food and Supplement Industries," *Marine Drugs* 13, no. 10 (2015).
16. Rosalee S. Rasmussen and Michael T. Morrissey, "Marine Biotechnology for Production of Food Ingredients," in *Advances in Food and Nutrition Research* (Academic Press, 2007).
17. Ioannis S. Boziaris, "Food Ingredients from the Marine Environment. Marine Biotechnology Meets Food Science and Technology," *Frontiers in Marine Science* 1, no. 66 (2014).
18. Suleria et al., "Marine-Based Nutraceuticals: An Innovative Trend in the Food and Supplement Industries."
19. Nordic Innovation, "Novel Bioactive Seaweed Based Ingredients and Products," https://www.nordicinnovation.org/programs/novel-bioactive-seaweed-based-ingredients-and-products.
20. Council of the European Union European Parliament, "Regulation (Ec) No 1223/2009 of the European Parliament and of the Council of 30 November 2009 on Cosmetic Products (Text with Eea Relevance)," (2009), https://eur-lex.europa.eu/legal-content/EN/ALL/?uri=CELEX%3A32009R1223.
21. Jean-Baptiste Guillerme, Céline Couteau, and Laurence Coiffard, "Applications for Marine Resources in Cosmetics," *Cosmetics* 4, no. 3 (2017).
22. E. G. Brunt and J. G. Burgess, "The Promise of Marine Molecules as Cosmetic Active Ingredients," *International Journal of Cosmetic Science* 40, no. 1 (2018).
23. Guillerme, Couteau, and Coiffard, "Applications for Marine Resources in Cosmetics."

24. Brunt and Burgess, "The Promise of Marine Molecules as Cosmetic Active Ingredients."
25. Guillerme, Couteau, and Coiffard, "Applications for Marine Resources in Cosmetics."
26. Brunt and Burgess, "The Promise of Marine Molecules as Cosmetic Active Ingredients."
27. Guillerme, Couteau, and Coiffard, "Applications for Marine Resources in Cosmetics."
28. Brunt and Burgess, "The Promise of Marine Molecules as Cosmetic Active Ingredients."
29. Pôle Mer Bretagne Atlantique, "Maritime Economy," https://www.pole-mer-bretagne-atlantique.com/en/cluster/ecosysteme/maritime-economy.
30. Sathianeson Satheesh, Mohammad Abdulaziz Ba-akdah, and Abdulmohsin A. Al-Sofyani, "Natural Antifouling Compound Production by Microbes Associated with Marine Macroorganisms — A Review," *Electronic Journal of Biotechnology* 21 (2016).
31. Sivan Nir and Meital Reches, "Bio-Inspired Antifouling Approaches: The Quest Towards Non-Toxic and Non-Biocidal Materials," *Current Opinion in Biotechnology* 39 (2016).
32. Ronald M. Atlas and Terry C. Hazen, "Oil Biodegradation and Bioremediation: A Tale of the Two Worst Spills in U.S. History," *Environmental science & technology* 45, no. 16 (2011).
33. Mahsa Baniasadi and Seyyed Mousavi, "A Comprehensive Review on the Bioremediation of Oil Spills," in *Microbial Action on Hydrocarbons*, ed. Kumar M. Kumar V., Prasad R. (Singapore: Springer, 2018).
34. Nedaa Ali et al., "Bioremediation of Soils Saturated with Spilled Crude Oil," *Scientific Reports* 10, no. 1 (2020).
35. Maria Nikolopoulou and Nicolas Kalogerakis, "Biostimulation Strategies for Fresh and Chronically Polluted Marine Environments with Petroleum Hydrocarbons," *Journal of Chemical Technology & Biotechnology* 84, no. 6 (2009).
36. Nature Research, "Biosensors," https://www.nature.com/subjects/biosensors#:~:text=Biosensors%20are%20devices%20used%20to,transducer%2C%20and%20a%20reader%20device.

37. Silke Kröger and Robin J. Law, "Biosensors for Marine Applications: We All Need the Sea, but Does the Sea Need Biosensors?," *Biosensors and Bioelectronics* 20, no. 10 (2005).
38. Flanders Marine Institute, "Marine Environmental in Situ Assessment and Monitoring Tool Box," http://www.vliz.be/en/imis?module=project&proid=4805
39. Oilgae, "Oilgae Guide to Fuels from Macroalgae," (2016), https://arpa-e.energy.gov/sites/default/files/Oilgae%20Guide%20to%20Fuels%20from%20Macroalgae%202010.pdf.
40. US Department of Energy, "Powering the Blue Economy: Exploring Opportunities for Marine Renewable Energy in Martime Markets," (2019), https://www.energy.gov/eere/water/powering-blue-economy-exploring-opportunities-marine-renewable-energy-maritime-markets.
41. Muhammad Imran Khan, Jin Hyuk Shin, and Jong Deog Kim, "The Promising Future of Microalgae: Current Status, Challenges, and Optimization of a Sustainable and Renewable Industry for Biofuels, Feed, and Other Products," *Microbial cell factories* 17, no. 1 (2018).
42. Abo Bodjui Olivier et al., "Microalgae to Biofuels Production: A Review on Cultivation, Application and Renewable Energy," *Reviews on Environmental Health* 34, no. 1 (2019).
43. Renganathan Rajkumar, Zahira Yaakob, and Mohd Sobri Takriff, "Potential of Micro and Macro Algae for Biofuel Production: A Brief Review," *2013* 9, no. 1 (2013).
44. Meiron Zollmann et al., "Green Technology in Green Macroalgal Biorefineries," *Phycologia* 58, no. 5 (2019).
45. Mustafa Vohra et al., "Bioethanol Production: Feedstock and Current Technologies," *Journal of Environmental Chemical Engineering* 2, no. 1 (2014).
46. John J. Milledge and Patricia J. Harvey, "Potential Process 'Hurdles' in the Use of Macroalgae as Feedstock for Biofuel Production in the British Isles," *Journal of chemical technology and biotechnology (Oxford, Oxfordshire: 1986)* 91, no. 8 (2016).
47. Rajkumar, Yaakob, and Takriff, "Potential of Micro and Macro Algae for Biofuel Production: A Brief Review."

48. Oilgae, "Oilgae Guide to Fuels from Macroalgae."
49. Ltd. Mitsui O.S.K. Lines, "Mol and Euglena Co. Conduct Sea Trial of Mol Group Tugboat Using Euglena Biodiesel Fuel for the First Time in Port of Nagoya," (2020), https://www.mol.co.jp/en/pr/2020/img/20088.pdf.
50. Khan, Shin, and Kim, "The Promising Future of Microalgae: Current Status, Challenges, and Optimization of a Sustainable and Renewable Industry for Biofuels, Feed, and Other Products."
51. US Department of Energy, "Powering the Blue Economy: Exploring Opportunities for Marine Renewable Energy in Martime Markets"
52. Martín P. Caporgno and Alexander Mathys, "Trends in Microalgae Incorporation into Innovative Food Products with Potential Health Benefits," *Frontiers in nutrition* 5 (2018).
53. Khan, Shin, and Kim, "The Promising Future of Microalgae: Current Status, Challenges, and Optimization of a Sustainable and Renewable Industry for Biofuels, Feed, and Other Products."
54. Oilgae, "Oilgae Guide to Fuels from Macroalgae"

References

Ali, Nedaa, Narjes Dashti, Majida Khanafer, Husain Al-Awadhi, and Samir Radwan. "Bioremediation of Soils Saturated with Spilled Crude Oil." *Scientific Reports* 10, no. 1 (2020/01/24/ 2020): 1116.

Assefa, Ayalew, and Fufa Abunna. "Maintenance of Fish Health in Aquaculture: Review of Epidemiological Approaches for Prevention and Control of Infectious Disease of Fish." *Veterinary Medicine International* 2018 (2018/02/26/ 2018): 5432497.

Atlas, Ronald M., and Terry C. Hazen. "Oil Biodegradation and Bioremediation: A Tale of the Two Worst Spills in US History." [In eng]. *Environmental science & technology* 45, no. 16 (2011): 6709–15.

Baniasadi, Mahsa, and Seyyed Mousavi. "A Comprehensive Review on the Bioremediation of Oil Spills." In *Microbial Action on Hydrocarbons*, edited by Kumar M. Kumar V., Prasad R., 223–54. Singapore: Springer, 2018.

Bodjui Olivier, Abo, Odey Emmanuel Alepu, Bakayoko Moussa, and Kalakodio Loissi. "Microalgae to Biofuels Production: A Review on Cultivation, Application and Renewable Energy." [In English]. *Reviews on Environmental Health* 34, no. 1 (2019/01/03/ 2019): 91–99.

Boziaris, Ioannis S. "Food Ingredients from the Marine Environment. Marine Biotechnology Meets Food Science and Technology." [In English]. *Frontiers in Marine Science* 1, no. 66 (2014/11/21/ 2014).

Brunt, E. G., and J. G. Burgess. "The Promise of Marine Molecules as Cosmetic Active Ingredients." *International Journal of Cosmetic Science* 40, no. 1 (2018/02/01/ 2018): 1–15.

Caporgno, Martín P., and Alexander Mathys. "Trends in Microalgae Incorporation into Innovative Food Products with Potential Health Benefits." [In eng]. *Frontiers in nutrition* 5 (2018): 58–58.

Council, NR, ER Commission on Geosciences, OS Board, and CORH Health. *From Monsoons to Microbes: Understanding the Ocean's Role in Human Health.* National Academies Press, 1999.

Dadar, Maryam, Kuldeep Dhama, Vikram N. Vakharia, Seyed Hossein Hoseinifar, Kumaragurubaran Karthik, Ruchi Tiwari, Rekha Khandia, *et al.* "Advances in Aquaculture Vaccines against Fish Pathogens: Global Status and Current Trends." *Reviews in Fisheries Science & Aquaculture* 25, no. 3 (2017/07/03/ 2017): 184–217.

European Parliament, Council of the European Union. "Regulation (Ec) No 1223/2009 of the European Parliament and of the Council of 30 November 2009 on Cosmetic Products (Text with Eea Relevance)." (2009). https://eur-lex.europa.eu/legal-content/EN/ALL/?uri=CELEX%3A32009R1223.

Flanders Marine Institute. "Marine Environmental in Situ Assessment and Monitoring Tool Box." http://www.vliz.be/en/imis?module=project&proid=4805

Gov, Esra, and Kazim Y. Arga. "Systems Biology Solutions to Challenges in Marine Biotechnology." [In English]. *Frontiers in Marine Science* 1, no. 14 (2014/06/20/ 2014).

Guillerme, Jean-Baptiste, Céline Couteau, and Laurence Coiffard. "Applications for Marine Resources in Cosmetics." *Cosmetics* 4, no. 3 (2017).

Khan, Muhammad Imran, Jin Hyuk Shin, and Jong Deog Kim. "The Promising Future of Microalgae: Current Status, Challenges, and Optimisation of a Sustainable and Renewable Industry for Biofuels, Feed, and Other Products." [In eng]. *Microbial cell factories* 17, no. 1 (2018): 36–36.

Kröger, Silke, and Robin J. Law. "Biosensors for Marine Applications: We All Need the Sea, but Does the Sea Need Biosensors?" *Biosensors and Bioelectronics* 20, no. 10 (2005/04/15/ 2005): 1903–13.
Lindequist, Ulrike. "Marine-Derived Pharmaceuticals - Challenges and Opportunities." [In eng]. *Biomolecules & therapeutics* 24, no. 6 (2016): 561–571.
Ma, Jie, Timothy J. Bruce, Evan M. Jones, and Kenneth D. Cain. "A Review of Fish Vaccine Development Strategies: Conventional Methods and Modern Biotechnological Approaches." [In eng]. *Microorganisms* 7, no. 11 (2019): 569.
Milledge, John J., and Patricia J. Harvey. "Potential Process 'Hurdles' in the Use of Macroalgae as Feedstock for Biofuel Production in the British Isles." [In eng]. *Journal of chemical technology and biotechnology (Oxford, Oxfordshire : 1986)* 91, no. 8 (2016): 2221–34.
Mitsui OSK. Lines, Ltd. "Mol and Euglena Co. Conduct Sea Trial of Mol Group Tugboat Using Euglena Biodiesel Fuel for the First Time in Port of Nagoya." (2020). https://www.mol.co.jp/en/pr/2020/img/20088.pdf
Mohd-Aris, Aslizah, Mohd Hafiz Ngoo Muhamad-Sofie, Mohd Zamri-Saad, Hassan Mohd Daud, and Md Yasin Ina-Salwany. "Live Vaccines against Bacterial Fish Diseases: A Review." [In eng]. *Veterinary world* 12, no. 11 (2019): 1806–15.
Nature Research. "Biosensors." https://www.nature.com/subjects/biosensors#:~:text=Biosensors%20are%20devices%20used%20to,transducer%2C%20and%20a%20reader%20device.
Nikolopoulou, Maria, and Nicolas Kalogerakis. "Biostimulation Strategies for Fresh and Chronically Polluted Marine Environments with Petroleum Hydrocarbons." *Journal of Chemical Technology & Biotechnology* 84, no. 6 (2009/06/01/ 2009): 802–07.
Nir, Sivan, and Meital Reches. "Bio-Inspired Antifouling Approaches: The Quest Towards Non-Toxic and Non-Biocidal Materials." *Current Opinion in Biotechnology* 39 (2016/06/01/ 2016): 48–55.
Nordic Innovation. "Novel Bioactive Seaweed Based Ingredients and Products." https://www.nordicinnovation.org/programs/novel-bioactive-seaweed-based-ingredients-and-products.
OECD. "Marine Biotechnology: Enabling Solutions for Ocean Productivity and Sustainability." (2013). https://www.oecd.org/greengrowth/marine-biotechnology-ocean-productivity-sustainability.htm.
Oilgae. "Oilgae Guide to Fuels from Macroalgae." (2016). https://arpa-e.energy.gov/sites/default/files/Oilgae%20Guide%20to%20Fuels%20from%20Macroalgae%202010.pdf.

PHG Foundation. "Rna Vaccines: An Introduction." (2018). https://www.phgfoundation.org/briefing/rna-vaccines.

Pôle Mer Bretagne Atlantique. "Maritime Economy." https://www.pole-mer-bretagne-atlantique.com/en/cluster/ecosysteme/maritime-economy.

Rajkumar, Renganathan, Zahira Yaakob, and Mohd Sobri Takriff. "Potential of Micro and Macro Algae for Biofuel Production: A Brief Review." *2013* 9, no. 1 (2013/12/02/ 2013): 28.

Rasmussen, Rosalee S., and Michael T. Morrissey. "Marine Biotechnology for Production of Food Ingredients." In *Advances in Food and Nutrition Research*, 237–92: Academic Press, 2007.

Satheesh, Sathianeson, Mohammad Abdulaziz Ba-akdah, and Abdulmohsin A. Al-Sofyani. "Natural Antifouling Compound Production by Microbes Associated with Marine Macroorganisms — A Review." *Electronic Journal of Biotechnology* 21 (2016/05/01/ 2016): 26–35.

Suleria, Hafiz A., Simone Osborne, Paul Masci, and Glenda Gobe. "Marine-Based Nutraceuticals: An Innovative Trend in the Food and Supplement Industries." *Marine Drugs* 13, no. 10 (2015).

University of Aberdeen. "Development of Novel Oral Vaccination Strategies for Atlantic Salmon." https://www.abdn.ac.uk/sfirc/research/vaccine-development/noval-oral-vaccination-strategies/.

US Department of Energy. "Powering the Blue Economy: Exploring Opportunities for Marine Renewable Energy in Martime Markets." (2019). https://www.energy.gov/eere/water/powering-blue-economy-exploring-opportunities-marine-renewable-energy-maritime-markets.

Vohra, Mustafa, Jagdish Manwar, Rahul Manmode, Satish Padgilwar, and Sanjay Patil. "Bioethanol Production: Feedstock and Current Technologies." *Journal of Environmental Chemical Engineering* 2, no. 1 (2014/03/01/ 2014): 573–84.

Zollmann, Meiron, Arthur Robin, Meghanath Prabhu, Mark Polikovsky, Amichai Gillis, Semion Greiserman, and Alexander Golberg. "Green Technology in Green Macroalgal Biorefineries." *Phycologia* 58, no. 5 (2019/09/03/ 2019): 516–34.

7
Marine Renewable Energy

Introduction

Globally, marine renewable energy (MRE) programmes are being implemented to mitigate carbon emissions, address the potential future exhaustion of fossil fuel supplies, and help ensure national energy security.[1] The main types of MRE systems are offshore wind energy and ocean energy (sometimes referred to as Blue Energy), which comprises energy from waves, tides/sea currents, and thermal and salinity gradients.[2] This chapter will first discuss offshore wind energy and related issues before discussing wave and tidal current energy, ocean thermal energy conversion, and salinity gradient energy. The chapter will then provide an overview of the environmental impacts of MRE before introducing the concept of the Environmental Impact Assessment (EIA) and marine spatial planning (MSP) in the context of MRE.

R. C. Brears, *Developing the Blue Economy*,
https://doi.org/10.1007/978-3-030-84216-1_7

Offshore Wind Energy

Wind is the result of temperature differences in different places—uneven heating results in a difference in atmospheric pressures, which causes the air to move. The kinetic energy of the moving air (wind) is transformed into electrical energy by wind turbines or wind energy conversion systems. Wind power is proportional to the rotor's dimensions and the cube of the wind speed: theoretically, when wind speed doubles, the wind power increases eight times.[3]

Offshore wind energy is the most advanced form of MRE in terms of technology development, policy frameworks, and installed capacity. Winds are usually stronger and more stable at sea than on land, resulting in more significant electricity generation per unit installed and a more consistent power generation pattern. Offshore wind turbines can also be larger than land-based turbines, therefore, generating more power per turbine.[4] Other benefits of offshore wind energy include:

- A lower visual and acoustic impact than wind farms on land, allowing better use of existing wind resources with larger turbines and the use of more efficient blade geometries. Also, the lower surface roughness of the sea favours the use of lower tower heights
- Creating more employment opportunities in the phases of construction, assembly, and maintenance due to the greater complexity during installation and exploitation[5,6]

Offshore Wind Foundations

The difference between onshore and offshore wind farms (a power generating facility that contains a number of wind turbines) is the foundation. Onshore wind turbines stand on concrete foundations, while offshore turbines have their foundations on the sea bed (fixed-bottom) or in the water (floating). Each foundation type is designed to support the wind turbine portions above sea level (the wind turbine tower, the nacelle that contains the generator, the rotor hub, and the turbine blades). The foundation needs to be able to resist two types of stressors:

- Vertical loads from the weight of the wind turbine component
- Horizontal loads from the force of the winds, ocean currents, and waves

Fixed-Bottom Foundations

Fixed-bottom foundations vary in types of structures, as detailed below, while Table 7.1 summarises the differences between each structure. Nonetheless, fixed-bottom foundations can only be used in shallow water (up to 30 metres in depth):

- *High-rise pile cap*: Steel piles are anchored in the seabed, and a pile cap is poured to connect with the pile top. The entire turbine can be assembled on land and transported by a purpose-built barge to the site for final erection
- *Monopile*: This is a simple design where the monopile supports the tower, either directly or through a transition piece, which is a transitional section between the tower and the monopile. The monopile continues down into the soil. The structure is made of cylindrical steel tubes
- *Gravity-based structure*: This concrete-based structure can be constructed with or without small steel or concrete skirts. The base width is adjusted for actual soil conditions. The design includes a central steel or concrete shaft for transition to the wind turbine tower
- *Tripod*: Made of cylindrical steel tubes, the central steel shaft of the tripod makes the transition to the wind turbine tower. The base width and pile penetration depth can be adjusted to suit the actual environmental and soil conditions
- *Tri-pile*: This consists of three cylindrical pile legs that connect to a transition piece above the waterline, forming a space frame that supports the wind turbine tower and turbine
- *Jacket*: Having four piles instead of three, the metal piles are linked together with a lattice that provides strength and stability to the whole structure
- *Suction bucket*: These foundations, normally made from steel or concrete, are installed using the principles of suction, whereby the

Table 7.1 Different Fixed-bottom Foundations Available for Offshore Wind Turbines

Type	Depth (m)	Pros	Cons	Comments
High-rise pile cap	0–20	Cap protects against marine collisions	• Limited water depth • Complex manufacturing	Standard in the onshore industry
Monopile	0–30	Simple design	• The diameter increases significantly with depth • Drilling	• Most widespread foundation type • Limitations in water depth
Gravity base	0–40	Cheap, no drilling	Seabed preparation required	Currently only used in shallow water
Tripod	0–49	More stable than monopile	More complex installation	High production costs due to complex structure and weight
Tri-pile	0–50	It can be installed by traditional jack-up barge	Cost	High production costs due to a complex structure and weight
Jacket	0–50	Stable and light	Cost	Commercially attractive due to their flexibility and low weight (40–50% less steel than monopiles)
Suction bucket	0–55	Less steel, no drilling required	Not applicable to the hard seabed	Yet to be deployed at scale

pressure difference generated between the inside of the bucket and the water surrounding it (at the seabed) leads to the structure being installed without any use of mechanical force

Floating Foundations

These are usually used at depths exceeding 50–60 metres because the cost of fixed-bottom foundations is prohibitive in deeper waters. There are three types of floating foundations in terms of how the design achieves its stability:

- *Ballast stabilised*: This uses a very large weight deep underwater, providing a counterbalance to the loads. Simple moorings are used to keep the structure in place
- *Mooring line stabilised (tension leg platform)*: This uses tensioned mooring arrangements to keep the structure stable
- *Buoyancy stabilised foundations*: This uses the waterplane area to achieve stability, like the way a barge does. Simple moorings are used to keep the structure in place[7,8,9,10,11,12]

Case: Offshore Wind Energy in Maryland

The Maryland Energy Administration (MEA) provides funding opportunities in Fiscal Year 2021 for businesses, including minority-owned emerging businesses, non-profits, and state, local, and municipal governments, and their institutions, to prepare the state's business supply chain and workforce entering the offshore wind industry:

- The Maryland Offshore Wind Capital Expenditure Program, with a budget of $1.6 million, provides grant funding on a competitive basis to support new or existing businesses entering the offshore wind supply chain in Maryland. The funding is provided to offset capital expenditure investments with grants available to cover 50% of a successful applicant's total project costs. The applicant contributes the remainder as matching funds or in-kind services. Grants may not exceed $400,000 and can be used to offset capital expenditures such as business expansion, including relocation costs, new construction of real property, and renovations to existing property, and the purchase and installation of new or manufacturer-refurbished equipment at a new or existing real property

- The Maryland Offshore Wind Workforce Training Program for Fiscal Year 2021 provides grant funding on a competitive basis to support new or existing workforce training centres entering the offshore wind industry by offsetting their capital expenditure investments and operating expenses. With a budget of $1.2 million, MEA will provide grants of up to $400,000 to fund projects that provide education of a trade skill(s) and safety training for the offshore wind industry in Maryland. Applicants must contribute at least 25% of the total project costs as matching funds or in-kind services. The funds can be used to offset capital expenditures such as the creation or expansion of a training centre and/or the purchasing and installation of new or manufacturer-refurbished training-related equipment[13]

Case: Denmark's Energy Islands

Denmark is establishing two energy islands off its coast. The energy islands, which will be completed by 2030, will supply 5 GW of power from offshore wind farms. One energy island will be in the North Sea, and the other in the Baltic. VindØ (wind island) is the larger of the two and will be situated in the North Sea. It will be created with funding from Danish pension funds PFA and PensionDanmark, and the Danish utility Andel. The island will have a capacity of 3 GW of offshore wind. The energy island cost will amount to nearly EUR 8 billion, regardless of whether it is constructed as a caisson embankment or a steel platform. The price includes offshore wind farms and power transmission. Meanwhile, Bornholm Energy Island, with 2 GW of wind power capacity, will use the existing island of Bornholm as the site for energy generation in the Baltic Sea. It will connect Germany and Denmark to the offshore hub.[14,15]

Wave and Tidal Current Energy

Wave and tidal energy are the two types of ocean energy that are expected to contribute significantly to the future supply of energy.[16]

Wave Energy

Ocean surface waves are generated by the wind blowing across the ocean. Waves constitute a substantial energy resource with very few environmental impacts from the construction of wave energy facilities. Wave energy converters convert ocean waves into electricity, the various types of which are listed in Table 7.2. Wave energy is harvested using either floating or fixed constructions. Floating devices convert the wave energy by coupling it to a hydraulic system, with the device lifting up and down with the waves' movements. Fixed devices use the oscillating water column generated by the wave to push air through a turbine.[17,18,19] Nonetheless, generating electricity from waves is a challenge due to varying wave climate, wave direction and variability, water depth, currents, and distance from shore. Therefore, different sites and environmental conditions will require different optimal solutions.[20]

Case: King Island Wave Power Trial

The Australian Renewable Energy Agency has provided $4 million funding for a new $12.3 million wave power demonstration project in King Island, Tasmania. The funding will support a Melbourne-based company to design, construct, install, and operate a 200-kW wave energy converter. The wave converter, which will sit partially submerged on the seabed, will use oscillating water column technology to push air into a chamber filled with an electricity-generating turbine. Electricity produced by the unit will be fed into the King Island microgrid, which was built to provide a reliable and stable supply of renewable energy and reduce reliance on diesel generators. The consistency of wave energy reduces the need for significant battery storage. The unit will be connected to the grid through a Power Purchase Agreement with Hydro Tasmania and operate for at least 12 months.[21,22]

Table 7.2 Types of Wave Energy Converters

Type	Description
Attenuator	This is a floating device that operates parallel to the wave direction. These devices capture energy from the relative motion of the two arms as the wave passes them
Point absorber	This is a floating structure that absorbs energy from all directions through its movement at/near the water surface. It converts the movement of the buoyant top relative to the base into electrical power
Oscillating wave surge converter	These extract energy from wave surges and the movement of water particles within. The arm oscillates as a pendulum mounted on a pivot point joint in response to the movement of the waves
Oscillating water column	This is a partially submerged, hollow structure. It is open to the sea below the waterline, enclosing a column of air on top of a water column. Waves cause the water column to rise and fall, compressing and decompressing the air column. This trapped air flows to/from the atmosphere via a turbine that can rotate regardless of the airflow direction. The rotating turbine is used to generate electricity
Overtopping/terminator device	These devices capture water as waves break into a storage reservoir. The water is then returned to the sea passing through a low-head turbine which generates power

(continued)

Table 7.2 (continued)

Type	Description
Submerged pressure differential	These are located near shore and attached to the seabed. The waves' motion causes the sea level to rise and fall above the device, inducing a pressure differential in the device. The alternating pressure pumps fluid through a system to generate electricity
Bulge wave	This consists of a rubber tube filled with water, moored to the seabed heading into the waves. The water enters through the stern, and the passing wave causes pressure variations along the tube's length, creating a "bulge". As the bulge travels through the tube, it grows. The energy from this can be used to drive a low-head turbine at the bow, where the water returns to the sea
Rotating mass	The movement of the device heaving and swaying in the wave can be attached to an electric generator inside the device

Tidal Energy

Solar and lunar gravitational forces, combined with the Earth's rotation, generate periodic changes in sea level known as tides. The rise and fall of ocean waters can be amplified by basin resonance and coastline topography to create substantial surface elevation changes in specific geographic locations. Most locations have two high and two low tides per day. The difference between a high and low tide is called the tidal range, and it can vary each day depending on the sun's location and the moon and globally depending on the coastal location. The vertical rise and fall of water are supplemented by an incoming (flood) or outgoing (ebb) horizontal flow of water in bays, harbours, and estuaries, etc., which is called a tidal current. Tidal energy can be harnessed using:

- *Large barrages in high tidal range areas*: This approach involves a structure impounding a large tidal body of water. When the tidal height varies outside of the impounded area, water is discharged either into or out of the enclosed area through conventional hydro turbines housed in the structure, creating electrical power as the water moves from one side of the dam to the other

- *Tidal Lagoon*: This power station generates electricity from the tides' natural rise and fall. Tidal lagoons work similarly to tidal barrages. They capture a large volume of water behind a human-made structure that is then released to drive turbines and generate electricity. Unlike a barrage that spans an entire river estuary in a straight line, a tidal lagoon encloses an area of a coastline with a high tidal range behind a breakwater
- *Tidal fences*: Tidal fences are composed of a number of individual vertical axis turbines mounted within a fence structure. Unlike barrages, tidal fences can also be used in unconfined basins, for example, in the channel between the mainland and a nearby offshore island or between two islands. They can generate electricity once the initial modules are installed, rather than after complete installation as in the case of barrage technologies
- *Tidal turbines*: This is a hydrokinetic approach that utilises the same technology like wind turbines to convert the kinetic energy of flowing water into electricity using underwater turbines (Table 7.3). While the principles of energy extraction from a moving fluid are well-known, and the environmental impact of tidal turbines being lower than the barrage approach, there are a few factors that need to be considered when designing tidal turbines:

  - The density of water is around 850 times that of air, leading to a smaller required capture area but with higher associated loads
  - Power density increases with the cube of the flow velocity, with the power density of a three m/s flow more than triple that of a two m/s flow
  - Alternating flow direction with tidal flows reversing a few times per day
  - Minimal accessibility, with installation operations only being done at low current speeds, resulting in accessibility being as short as 30 minutes
  - Significant flow shear and turbulence levels vary spatially and temporally, impacting loading and performance[23,24,25,26,27]

Case: Scotland's Pentland Firth Tidal Energy Project

In 2010, as part of the Pentland Firth and Orkney Waters leasing round, The Crown Estate awarded an agreement for lease to MeyGen Limited, granting the option to develop a tidal stream project up to 398 MW. The first phase of the MeyGen project (Phase 1A) involved MeyGen installing four horizontal axis turbines at the Inner Sound site of the Pentland Firth. The turbines result in an overall capacity of 6 MW as part of MeyGen's "deploy and monitor strategy" and act as a precursor to developing the remaining consented 86 MW project. The project aims to demonstrate the commercial viability and technical feasibility of tidal array projects, with lessons learned from the construction, installation,

Table 7.3 Types of Tidal Turbines

Type	Description
Horizontal axis turbine	This extracts energy from moving water the same way wind turbines extract energy from moving air. The tidal stream causes the rotors to rotate around the horizontal axis and generate power
Vertical axis turbine	The turbine is mounted on a vertical axis. The tidal stream causes the rotors to rotate around the vertical axis and generate power
Oscillating hydrofoil	A hydrofoil is attached to an oscillating arm. The tidal current flowing on either side of the wing results in lift. This motion then drives fluid in a hydraulic system to be converted into electricity
Enclosed tips	A Venturi effect device in a duct concentrates tidal flow passing through the turbine. The tunnel-like device sits submerged in the tidal current. The flow of water can drive the turbine directly, or the induced pressure differential in the system can drive an air-turbine
Archimedes screw	This is a helical corkscrew-shaped device. The device draws power from the tidal stream as the water moves up/through the spiral turning the turbines
Tidal kite	A tidal kite is tethered to the sea bed and carries a turbine below the wing. The kite "flies" in the tidal stream, making a figure-of-eight shape to increase the speed of the water flowing through the turbine

operation, and maintenance of this phase of the project fed into subsequent phases. In 2019, the project sent nearly 14 GWh of electricity to the grid, almost double the previous high of 7.4 GWh in 2018. In total, the project has exported more than 25.5 GWh of electricity to the grid since 2017.[28,29]

Ocean Thermal Energy Conversion

Ocean thermal energy conversion (OTEC) is based on the concept that significant power can be generated by utilising differences between surface and subsurface ocean water temperature, in particular, using the temperature difference between warm seawater at the ocean's surface and cold seawater at between 800 and 1,000 metres depth to produce electricity. The warm seawater is used to produce a vapour that acts as a working fluid to drive a turbine/generator. There are four main types of OTEC:

- *Open cycle OTEC*: Warmer surface water is introduced through a valve in a low-pressure compartment and flash evaporated. The vapour drives a generator and is condensed by the cold seawater pumped up from below
- *Closed cycle OTEC*: Surface water, with higher temperatures, provides heat to a working fluid with a low boiling temperature, providing vapour pressure. Usually, ammonia is used as a working fluid. The vapour drives a generator that produces electricity. The working fluid vapour is then condensed by the cold water from the deep ocean and pumped back in a closed system
- *Kalina cycle OTEC*: This is a variation of a closed cycle OTEC where instead of pure ammonia, a mixture of water and ammonia is used as the working fluid. The mixture does not have a boiling point but instead has a boiling point trajectory. More of the provided heat is taken into the working fluid during evaporation, and therefore, more heat can be converted, and efficiencies are enhanced

- *Hybrid system*: These systems combine open and closed cycles where the steam generated by flash evaporation is then used as heat to drive a closed cycle:
  1. Electricity is generated in a closed cycle system, as discussed above
  2. The warm seawater discharges from the closed cycle OTEC are flash evaporated like an open cycle OTEC system and cooled with the cold water discharge[30,31,32]

Case: Puerto Rico Ocean Technology Complex
The Government of Puerto Rico, through the Department of Economic Development and Commerce, is issuing a Request for Proposals to prequalified respondents that have the expertise and ability to design, build, operate, maintain, and finance the Puerto Rico Ocean Technology Complex (PROTech), which includes an OTEC facility. The OTEC facility is planned to be the first large-scale plant globally, with a capacity of between 5 and 10 MW. The PROTech complex will also have a seawater air conditioning (SWAC) system installed to meet the Research Park's principal tenants' cooling demand. The outputs of the SWAC system provide a variety of opportunities to develop a range of products and services, including:

- Potable water, with desalination providing high-quality drinking water
- Aquaculture/mariculture development for growing fish/shrimps in clean ponds
- Aquaculture for growing exotic species for aquariums
- Algae cultivation for cosmetic, health/medical, and/or biofuels production[33]

Salinity Gradient Energy

Salient gradient power is the energy created from the difference between two fluids, commonly fresh and saltwater, where a river flows into the sea. The theoretical amount of energy available from mixing one cubic metre of seawater with one cubic metre of river water is 1.4 megajoules. There are two standard technologies used involving membranes:

- *Pressure Retarded Osmosis (PRO)*: This uses a membrane to separate a concentrated salt solution (like seawater) from freshwater. The freshwater flows through a semi-permeable membrane towards the seawater, which increases the pressure within a seawater chamber. A turbine is spun as the pressure is compensated and electricity generated
- *Reversed Electro Dialysis (RED)*: This uses the transport of (salt) ions through membranes. RED consists of a stack of alternating cathode and anode exchanging permselective membranes. The compartments between the membranes are alternatively filled with seawater and freshwater. The salinity gradient difference is the driving force in transporting ions that results in an electrical potential, which is then converted to electricity[34]

Case: Osmotic Power Plant on the Oslo Fjord of Norway

In 2009, Statkraft completed and put into operation one of the world's first prototype osmotic power plants at Tofte on the Oslo Fjord of Norway. A paper pulp factory's premises were leased to Statkraft to construct the prototype as it had access to fresh water and seawater. The prototype, which ran until 2014, produced 2–4 kW. It operated on the PRO process, which involves pumping seawater at 60–85% of the osmotic pressure against one side of semi-permeable membranes whose other side is exposed to freshwater. The prototype also tested and measured environmental challenges, such as measuring potential algae bloom related to brackish water discharge.[35,36]

Environmental Impacts of Marine Renewable Energy

While offshore MRE provides a range of economic, environmental, and social benefits, including job creation, reduction in the cost of most technologies involved, and higher levels of development from increased access to modern energy services, MRE technology can be detrimental to the environment due to:

- The alteration of benthic habitats and sediment transport or deposition by the construction activities and continuous presence of devices and structures
- Noise and electromagnetic fields resulting in deaths or changes in the behaviour of fish and mammals
- Interference of the movement, feeding, spawning, and migration path of fish, mammals, and birds
- The release of toxic chemicals from accidental spills or leaks, or the accumulation of metals or organic compounds
- The reduction of the velocity of marine currents and decreases in wave height from the extraction of wave or tidal energy

Due to the complexity of the marine environment and the various technologies used to harvest MRE, it is essential to classify environmental effects within a framework, as detailed in Table 7.4.[37,38,39]

Minimising the Impact of Marine Renewable Energy Sites

MRE sites should have a sparse biological community and predominantly comprise opportunistic, resilient, coloniser species to minimise the impact on existing flora and fauna. Soft sediment communities are typically low in diversity and have species that are adapted to naturally unstable habitat conditions. Locations used as a migratory or periodic habitat essential for life history completion should be avoided.[40]

Table 7.4 Environmental Effects and Impacts of Marine Renewable Energy Projects

Component	Description
Stressors: These are features that may induce environmental changes	• The physical presence of fixed and moving parts of the devices in the water and the air, including the introduction of material or substrate at the bed • Near- and far-field changes in the air and water pressure fields and the sediment dynamics (including changes in sediment distribution during construction) • Release of chemicals in the area from the equipment and vehicles linked to the activity and from seabed removal • Generation of sound, both above and underwater • Electromagnetic fields, produced by cables (during the operational phase) • Cumulative impacts of stressors from several large-scale projects and other human activities For each stressor, the development stage of the project (survey, construction, operation and maintenance, and decommissioning) should be considered, along with the timescale (duration and frequency) and spatial extent: both the timescale and spatial extent are project- and site-specific
Receptors: These are elements of the ecosystem that may/may not respond to the stressor	• Physical environment, which comprises the atmosphere and marine (wave and current) climates and the bed sediment (near-field and far-field) • Marine mammals and sea turtles • Pelagic habitat and communities, including planktonic and nektonic organisms (excluding marine mammals and sea turtles) • Benthic habitat and communities, including macrophytes, invertebrates, and vertebrates living in association with bed sediment • Marine birds, living or migrating near the project area • Water quality, measured on its physical and chemical properties

(continued)

Table 7.4 (continued)

Component	Description
Effects: This describes how receptors are affected by stressors but do not indicate magnitude or significance	The general effects of stressors upon receptors include: • Collision potential • Scouring • Seabed disruption • Hydrodynamic changes • Aerodynamic changes • Sediment dynamic changes • Hearing injuries • Site avoidance • Stress increase • Behavioural change • Sediment temperature increase • Pollution from dredging • Leaching • Spilling • Pollution from maintenance
Impacts: This deals with the severity, intensity, or duration of the effect and with its direction (i.e., positive or negative)	• Impacts are generally recognised when the effects induce changes in specific variables that are used to define the status of the concerned receptor • The impacts can be either direct or indirect • Impacts are likely to vary with location and season, the design and scale of both devices and arrays, and may be cumulative in time and with an increasing number of devices • Indicators can be used to determine if the effects are strong enough to induce impacts and if a response is required Negative impacts of MRE systems include: • *Flow alteration*: Wave and tidal changes can affect the regulation of marine ecosystems and natural energy fluxes • *Wave climates*: MRE systems can alter patterns and rates of shoreline erosion, deposition, and scour, and change beach morphology and coastal habitats • *Current patterns*: Tidal current energy extraction can modify current velocities and dynamics, sediment transport pathways, patterns and rates of erosion, deposition, and scour, turbidity, etc., impacting the location of primary production and the suitability of foraging habitats

(continued)

Table 7.4 (continued)

Component	Description
	• *Sediment and nutrient transport*: During operation, MRE systems can modify sedimentation patterns, impacting benthic communities from increased turbidity and decreasing light availability • *Biofouling and pollution risk*: MRE systems are designed to survive harsh conditions. Biofouling may increase sedimentation rates, resulting in eutrophication. During operation and maintenance activities, there are risks of pollution, including chemical spills and leaching of chemicals from antifouling paints • *Collision risk*: The deployment of MRE systems results in many species at risk of collision or entanglement with moving machinery or mooring lines. This can result in recoverable injury, permanent debilitation, or delayed or instant mortality. The presence of these systems can modify habitat use and migration patterns and alter survival or reproduction rates • *Underwater noise*: Hearing injuries and habitat loss can occur during the construction phase. Production of noise during operation may mask bioacoustics for communication and navigation of long-distance migrating whales and sea turtle • *Invasive species*: Newly constructed structures may serve as steppingstones for invasive species communities • *Electromagnetic fields*: The production of magnetic fields by cables can modify the behaviour of resident or migratory species that use the geomagnetic field for localisation and orientation Positive impacts of MRE systems include: • Exclusion of fishing activities, including trawling, within the project area, increasing fish stocks • Providing additional (or new) settlement surface/habitat for benthic organisms and fishes, increasing biodiversity

Environmental Impact Assessment

In most jurisdictions, developers pursuing MRE are required to undertake some form of EIA before deployment. An EIA presents evidence (baseline data gathered through site characterisation surveys, modelling, evaluations, etc.) of likely environmental impacts and is done through four steps:

1. Identification of the environmental changes which may result from the development and the receptors that may be affected
2. Evaluation of the exposure risk and sensitivity of the receptors
3. Evaluation of the impact significance in relation to the vulnerability and exposure risk of the receptors
4. Identification of mitigation measures for any significant impacts identified and evaluation of the likely residual impacts[41]

Environmental Monitoring Programme

An environmental monitoring programme (EMP) is one of the most important outputs of an EIA, with the main objectives of:

- Providing feedback and early warning of potential environmental damage
- Ensuring the impacts do not exceed legal standards
- Checking the implementation of mitigation measures in the manner described in an environmental statement

Use of Environmental Indicators

MRE projects are usually located at specific locations offshore due to grid and land access considerations, along with site-specifics regarding the resource (particularly for waves and currents). The development of these large-scale projects, along with other human activities offshore, creates environmental concerns. The evaluation of the environmental effects of MRE is difficult as the marine environment is a highly complex system where physical, chemical, and biological properties interact at various spatial and temporal scales. Environmental indicators generally reduce the complexity of a problem, or of many parameters, to a smaller number of key parameters that enable the description or quantification of the status and trends of entire or partial ecosystems. Indicators can then facilitate management decisions as they provide decision-makers with information about where, when, and how to act. They also communicate overall progress on stated goals and benchmarks. The use of environmental indicators to report the results of the EMP allows the incipient offshore MRE project to:

- Convey information to government and industries about environmental effects (negative and positive)
- Determine whether observed effects are acceptable or not through the upfront specification of thresholds and trigger levels
- Assess the effectiveness of mitigation measures
- Make comparisons with similar projects and with other human activities
- Communicate with other stakeholders, such as non-governmental organisations, the public, etc.[42]

Marine Spatial Planning

Unlike single-sector management of the past, MSP is "*an integrated planning framework that informs the spatial distribution of activities in and on the ocean in order to support current and future uses of ocean ecosystems and maintain the delivery of valuable ecosystem services for future generations in a way that meets ecological, economical, and social objectives*".[43] Usually, the MSP process results in preparing a comprehensive plan or policy document often referred to as the master plan, which describes the vision for the future spatial development of the particular marine area. The plan is long term and is prepared based on interactions with all relevant stakeholders and spatial data collected for a particular marine area. Table 7.5 details the rationale for actively involving stakeholders in

Table 7.5 Rationale for Actively Involving Stakeholders in Marine Spatial Planning

Rationale	Description
Ownership	Encourages ownership of the plan, engenders trust among stakeholders and decision-makers, and encourages voluntary compliance with rules and regulations
Understanding of complexities	It improves understanding of the spatial and temporal complexity and human influences of the marine management area
Shared understanding of challenges	It develops a mutual and shared understanding of the problems and challenges in the management area
Understanding of sector perceptions	It increases understanding of underlying (often sector-orientated) desires, perceptions, and interests that drive and/or prohibit integration of policies in the management area
Compatibility and conflict	It examines existing and potential compatibility and/or conflicts of multiple-use objectives of the management area
New options	It aids the generation of new options, consensus, and solutions that may not have been considered individually
Capacity-building	It expands and diversifies the capacity of the planning team through the inclusion of secondary and tertiary information, for example, local knowledge

MSP. The policy or master plan is usually implemented through detailed ocean zoning maps, which partition areas into different zones where certain activities are permitted, restricted, or prohibited.[44,45] The steps for implementing MSP are listed in Table 7.6.[46]

Table 7.6 Steps for Marine Spatial Planning

Step	Description	Activities
1. Planning	This involves talking with managers to determine priorities	• Define goals and objectives • Define the study area
2. Data evaluation	This involves assessing the data and identifying data gaps	• Data acquisition • Data gap identification • Data content and quality
3. Ecosystem characterisation	This involves describing the ecosystem patterns and processes, including human activities across the area of interest	• Socio-economic analysis • Oceanographic analysis • Biological analysis • Habitat analysis
4. Management applications	This involves working with managers to support specific management applications	• Planning for MRE • Managing marine resources • Minimising conflicts • Designing Marine Protected Areas (MPA)

Data Management and Geographic Information Systems

Data is an asset, with it being an essential part of the evidence necessary to evaluate results. The value of data increases as it is aggregated into collections and becomes available for reuse to address new and challenging questions. Nonetheless, data must be reliable, timely, and relevant. MSP datasets must be managed in a coordinated manner to maximise the integrity of data available while complying with relevant legislation, best practices guidelines, and licensing conditions. Doing so enables MSP decisions to be consistent, open, sustainable, and evidence-based.[47] Complex data should be presented in an accessible manner to all stakeholders. This can be done effectively through local workshops, where community stakeholders meet to determine their needs by consensus, or use of Geographic Information Systems (GIS) as a tool to show relevant information.[48]

Geographic Information Systems

MSP aims to minimise the conflicts among different ocean users and their adverse effects by allocating space and applying zoning for different uses. GIS can be applied in several ways in the process:

- *GIS tools can be used for delineation of features*: The state of features (biogeophysical, socio-economical, and governance) in a marine area is essential knowledge in an informative MSP process, and GIS can be an efficient tool for identifying, locating, and visualising the cover and spatial distribution of resources and uses in the form of maps
- *GIS tools can be used as indicators for assessing management performance*: Management needs to be assessed regarding their efficiency and effects during the planning and implementation phases. This covers the whole MSP cycle. GIS tools can be used as indicators of the effects management has on biophysical, socio-economical, and governance factors. These tools are used as indicators of the state compared in different areas or the change of state over time
- *GIS tools can combine several types of spatial data*: The Pressure Evaluation Matrix (PEM) process relates all uses and their effects in a marine area to the sensitivity of the habitats and species of that area. GIS can be combined with the PEM to visualise the relationship, i.e., by overlaying the spatial distribution of pressure factors and sensitive habitats[49]

Benefits of Marine Spatial Planning

MSP can have significant environmental, economic, and social benefits when appropriately developed, including the following listed in Table 7.7.[50]

Marine Renewable Energy and Marine Spatial Planning

MRE has specific requirements from a planning process perspective. For example, MRE needs to be linked with other infrastructures such as grid

Table 7.7 Benefits of Marine Spatial Planning

Benefit	Description
Environmental	• Identification of biologically and ecologically important areas • Biodiversity objectives incorporated into planned decision-making • Identification and reduction of conflicts between humans and nature • Allocation of space for biodiversity and nature conservation • Establishment of the context for planning MPAs • Identification and reduction of the cumulative effects of human activities on marine ecosystems
Economic	• Greater certainty of access to desirable areas for new private sector investment • Identification of compatible uses within the same area of development • Reduction of conflict between incompatible uses • Improved capacity for new and changing human activities, including emerging technologies • Better safety during operation of human activities • Promotion of the efficient use of resources and space • Streamlining and transparency in the permit and licensing procedures
Social	• Improved opportunities for community and citizen participation • Identification of impacts of decisions on the allocation of ocean space (e.g., closure areas for specific uses, protected areas) for communities and economies onshore (e.g., employment, distribution of income) • Identification and improved protection of cultural heritage • Identification and preservation of social and spiritual values related to ocean

provision and access to ports. Also, MRE impacts the marine environment during the construction and decommissioning phases. During both phases, local destruction of marine habitats may be caused by installing or removing the related infrastructures, for example, wind turbine, platform, etc. Increased turbidity, noise, and vibrations may also affect the distribution of fish populations and marine mammals.

Furthermore, as a relatively new sector, MRE can also overlap with more traditional uses such as fishing and navigation. When multiple-use situations arise, it can be challenging to address multiple users' different interests and needs in mutually satisfying ways. This is made more difficult because compatibility between users and activities depends on oceanographic conditions, such as sea turbulence, the nature of the seabed, the size of the water column, and each project's size and characteristic.

As such, MSP needs to accurately identify MRE locations where there are neither conflicts with other activities or severe threats to biodiversity and the marine environment (sensitive habitats, migratory pathways, feeding, spawning, nursery grounds, etc.). This, in turn, provides certainty and confidence to developers and investors. A further benefit is that synergies can be developed between MRE and other uses and between the different types of energy production at sea. For example, the combination of aquaculture and wind farms when the substructure of wind farms located close to the coast can be used for aquaculture purposes. Synergies can also be developed between wave energy farms and shore defence: Wave energy devices help break the waves and limit damage to coastal installations. Anchoring systems could also be shared, for example, with tidal energy devices.

Meanwhile, there is the possibility that offshore wind farms could increase local biomass. Therefore, these sites could be developed in synergy with protected areas to regenerate fish stocks. To achieve these synergies, it is vital that all long-term options for multiple potential uses are presented at an early stage in the planning process and discussed thoroughly with stakeholders as adding other uses within existing situations or after the realisation of a project is usually difficult.[51]

Areas Available for Marine Renewable Energy Development

MSP is often interpreted to be synonymous with ocean zoning, which sets areas where MRE installation is directly excluded. This is generally in sites where other uses are already established. The alternative is an inclusive or a policy-based approach in which areas for MRE development are not excluded a priori. Instead, MSP can be used to advocate preferred activities or priorities, reflecting national policy objectives, for example:

- Ireland has not identified areas as being prohibited for MRE activities
- Portugal's regulations for certain activities create exclusion areas and safety zones
- In Sweden, no areas are entirely prohibited for MRE development, but additional licensing requirements may be needed in areas designated for conservation purposes
- In Scotland, preferred zones and locations are being developed
- In Japan, designated demonstration sites for MRE research and development have been selected by local governments proposing a demonstration site
- In the United States, preferred areas for offshore wind development have been designated in the Atlantic by the Bureau of Ocean Energy Management[52,53,54,55,56]

Case: Marine Spatial Planning for Offshore Wind Farms in Denmark
Denmark's spatial planning committee for offshore wind was established in 1995. The Danish Energy Agency leads the committee. It consists of government authorities responsible for the natural environment, safety at sea and navigation, offshore resources extraction, visual interests, and grid transmission conditions. The committee is bolstered with technical expertise in the areas mentioned above. The committee examines the engineering, economic, and planning options for offshore wind farms. The committee regularly assesses offshore wind farms' siting regarding other interests at sea and appropriate land uses. The committee is tasked with finding appropriate sites for offshore wind farms, specifically sites where the impact on nature and other sea uses is expected to be low while suitable for offshore harvesting of wind. When these sites are found, they are reserved for the establishment of offshore wind farms. The committee uses GIS mapping, with each government authority having

their GIS map and associated reserved areas outlined, for example, sailing routes and environmental protection sites, etc. When the GIS maps for each government authority are collated on top of each other, a precise picture forms of areas with no reservations. These areas are then evaluated concerning distance to shore, wind speeds, water depth, etc. Suggested sites are then discussed with the remaining marine authorities and the affected municipalities onshore. Once the public authorities agree to the placements of offshore wind farms, public hearings will commence.[57]

Notes

1. Alistair G. L. Borthwick, "Marine Renewable Energy Seascape," *Engineering* 2, no. 1 (2016).
2. Takvor H. Soukissian et al., "Marine Renewable Energy Clustering in the Mediterranean Sea: The Case of Pelagos Project," *Frontiers in Energy Research* 7, no. 16 (2019).
3. IRENA, "Wind Power Technology Brief," (2016), https://www.irena.org/-/media/Files/IRENA/Agency/Publication/2016/IRENA-ETSAP_Tech_Brief_Wind_Power_E07.pdf.
4. Joseph Appiott, Amardeep Dhanju, and Biliana Cicin-Sain, "Encouraging Renewable Energy in the Offshore Environment," *Ocean & Coastal Management* 90 (2014).
5. Sergio Sánchez et al., "Foundations in Offshore Wind Farms: Evolution, Characteristics and Range of Use. Analysis of Main Dimensional Parameters in Monopile Foundations," *Journal of Marine Science and Engineering* 7, no. 12 (2019).
6. Appiott, Dhanju, and Cicin-Sain, "Encouraging Renewable Energy in the Offshore Environment."
7. Bureau of Ocean Energy Management, "Comparison of Environmental Effects from Different Offshore Wind Turbine Foundations," (2020), https://www.boem.gov/sites/default/files/documents/environment/Wind-Turbine-Foundations-White%20Paper-Final-White-Paper.pdf.

8. Sánchez et al., "Foundations in Offshore Wind Farms: Evolution, Characteristics and Range of Use. Analysis of Main Dimensional Parameters in Monopile Foundations."
9. Orsted, "Our Experience with Suction Buckets Jacket," https://orsted.com/en/our-business/offshore-wind/wind-technology/suction-bucket-jacket-foundations.
10. IRENA, "Wind Power Technology Brief".
11. Center for Wind Energy at James Madison University, "Review of Options for Offshore Foundation Substructures," (2012), https://static1.squarespace.com/static/5b3cf461d274cb109aa488d8/t/5b8b43f370a6adae1d2abc0c/1535853556282/Offshore+Foundations+Substructure+report.pdf.
12. Xiaojing Sun and Diangui Huang, "Wind Power Development in Shanghai, China," *International Journal of Sustainable Energy* 33, no. 1 (2014).
13. Maryland Energy Administration, "Offshore Wind Energy in Maryland," https://energy.maryland.gov/Pages/Info/renewable/offshorewind.aspx.
14. State of Green, "This Is What the World's First Energy Island May Look Like," https://stateofgreen.com/en/partners/state-of-green/news/this-is-what-the-worlds-first-energy-island-may-look-like/.
15. "Bornholm Can Become the World's First Energy Island," https://stateofgreen.com/en/partners/orsted-a-global-leader-within-green-energy/solutions/bornholm-can-become-the-worlds-first-energy-island/#:~:text=In%20November%202019%2C%20Ørsted%20presented,the%20world's%20first%20energy%20island.
16. Andreas Uihlein and Davide Magagna, "Wave and Tidal Current Energy—A Review of the Current State of Research Beyond Technology," *Renewable and Sustainable Energy Reviews* 58 (2016).
17. Appiott, Dhanju, and Cicin-Sain, "Encouraging Renewable Energy in the Offshore Environment."
18. Lawrence E. Jones, *Renewable Energy Integration: Practical Management of Variability, Uncertainty, and Flexibility in Power Grids* (San Diego, CA: Elsevier Science & Technology, 2014).
19. European Marine Energy Centre, "Wave Devices," http://www.emec.org.uk/marine-energy/wave-devices/.

20. Malin Göteman et al., "Advances and Challenges in Wave Energy Park Optimization—A Review," *Frontiers in Energy Research* 8, no. 26 (2020).
21. ARENA, "Uniwave200 King Island Project—Wave Swell," https://arena.gov.au/projects/uniwave200-king-island-project-wave-swell/.
22. "Full Steam Ahead for King Island Wave Power Trial," https://arena.gov.au/blog/king-island-wave-power/.
23. Jones, *Renewable Energy Integration: Practical Management of Variability, Uncertainty, and Flexibility in Power Grids.*
24. Appiott, Dhanju, and Cicin-Sain, "Encouraging Renewable Energy in the Offshore Environment."
25. Tidal Lagoon Power, "What Is a Tidal Lagoon?," http://www.tidallagoonpower.com/tidal-technology/what-is-a-tidal-lagoon/#:~:text=A%20tidal%20lagoon%20is%20a,drive%20turbines%20and%20generate%20electricity.
26. Murdoch University, "Tidal Barrage & Tidal Turbines," http://www.see.murdoch.edu.au/resources/info/Tech/tidal/
27. European Marine Energy Centre, "Tidal Devices," http://www.emec.org.uk/marine-energy/tidal-devices/.
28. SIMEC Atlantis Energy, "Meygen," https://simecatlantis.com/projects/meygen/.
29. CNBC, "A Tidal Project in Scottish Waters Just Generated Enough Electricity to Power Nearly 4,000 Homes," https://www.cnbc.com/2020/01/27/tidal-project-generates-electricity-to-power-nearly-4000-homes.html.
30. Appiott, Dhanju, and Cicin-Sain, "Encouraging Renewable Energy in the Offshore Environment."
31. S. M. Masutani and P. K. Takahashi, "Ocean Thermal Energy Conversion (Otec)*," in *Encyclopedia of Ocean Sciences (Second Edition)*, ed. John H. Steele (Oxford: Academic Press, 2001).
32. IRENA, "Ocean Thermal Energy Conversion: Technology Brief," (2014), https://www.irena.org/publications/2014/Jun/Ocean-Thermal-Energy-Conversion.
33. DEDC, "Puerto Rico Ocean Technology Complex (Protech)," https://www.ddec.pr.gov/en/protech-rfp/

34. IRENA, "Salinity Gradient Energy Policy Brief," (2014), https://www.irena.org/-/media/Files/IRENA/Agency/Publication/2014/Jun/Salinity_Energy_v4_WEB.pdf.
35. Power, "Statkraft Shelves Osmotic Power Project," https://www.powermag.com/statkraft-shelves-osmotic-power-project/.
36. Øystein Skråmestø Sandvik and Stein Erik Skilhagen Simen Bræin, "Osmotic Power. From Prototype to Industry—What Will It Take?" in *3rd International Conference on Ocean Energy* (Bilboa 2010).
37. Erwan Garel et al., "Applicability of the "Frame of Reference" Approach for Environmental Monitoring of Offshore Renewable Energy Projects," *Journal of Environmental Management* 141 (2014).
38. Paul A. J. Bonar, Ian G. Bryden, and Alistair G. L. Borthwick, "Social and Ecological Impacts of Marine Energy Development," *Renewable and Sustainable Energy Reviews* 47 (2015).
39. Linus Hammar et al., "Introducing Ocean Energy Industries to a Busy Marine Environment," ibid., 74 (2017).
40. Andrew B. Gill, "Offshore Renewable Energy: Ecological Implications of Generating Electricity in the Coastal Zone," *Journal of Applied Ecology* 42, no. 4 (2005).
41. Clive J. Fox et al., "Challenges and Opportunities in Monitoring the Impacts of Tidal-Stream Energy Devices on Marine Vertebrates," *Renewable and Sustainable Energy Reviews* 81 (2018).
42. Garel et al., "Applicability of the "Frame of Reference" Approach for Environmental Monitoring of Offshore Renewable Energy Projects."
43. Melissa M. Foley et al., "Guiding Ecological Principles for Marine Spatial Planning," *Marine Policy* 34, no. 5 (2010).
44. Michaela Young, "Building the Blue Economy: The Role of Marine Spatial Planning in Facilitating Offshore Renewable Energy Development," *The International Journal of Marine and Coastal Law* 30, no. 1 (2015).
45. Sarah Twomey and Cathal O'Mahony, "Stakeholder Processes in Marine Spatial Planning: Ambitions and Realities from the European Atlantic Experience," in *Maritime Spatial Planning: Past, Present, Future*, ed. Jacek Zaucha and Kira Gee (Cham: Springer International Publishing, 2019).

46. OES-Environmental, "Oes-Environmental 2020 State of the Science Report: Environmental Effects of Marine Renewable Energy Development around the World," (2020), https://tethys.pnnl.gov/publications/state-of-the-science-2020.
47. Sarah Flynn et al., "Lessons from a Marine Spatial Planning Data Management Process for Ireland," *International Journal of Digital Earth* (2020).
48. Peter Becker, Gavin Burnell, and Tetsuzan Ron, "Using GIS to Improve Coastal Marine Spatial Planning," *Sea Technology* 53 (2012).
49. BALANCE, "GIS Tools for Marine Spatial Planning and Management." (2007), https://balance-eu.org/xpdf/balance-interim-report-no-28.pdf.
50. UNESCO, "Marine Spatial Planning: A Step-by-Step Approach toward Ecosystem-Based Management," (2009), http://www.mspglobal2030.org/wp-content/uploads/2019/03/Marine-spatial-planning-a-step-by-step-approach.pdf.
51. European Commission, "Energy Sectors and the Implementation of the Maritime Spatial Planning Directive," (2015), https://ec.europa.eu/maritimeaffairs/publications/energy-sectors-and-implementation-maritime-spatial-planning-directive_en.
52. OES-Environmental, "Oes-Environmental 2020 State of the Science Report: Environmental Effects of Marine Renewable Energy Development around the World."
53. Santiago Salvador, Luis Gimeno, and F. Javier Sanz Larruga, "The Influence of Maritime Spatial Planning on the Development of Marine Renewable Energies in Portugal and Spain: Legal Challenges and Opportunities," *Energy Policy* 128 (2019).
54. European Commission, "Energy Sectors and the Implementation of the Maritime Spatial Planning Directive."
55. Montserrat Abad Castelos, "Marine Renewable Energies: Opportunities, Law, and Management," *Ocean Development & International Law* 45, no. 2 (2014).
56. OES-Environmental, "Oes-Environmental 2020 State of the Science Report: Environmental Effects of Marine Renewable Energy Development around the World."

57. Danish Energy Agency, "Danish Experiences from Offshore Wind Development," (2015), https://ens.dk/sites/ens.dk/files/Globalcooperation/offshore_wind_development.pdf.

References

Abad Castelos, Montserrat. "Marine Renewable Energies: Opportunities, Law, and Management." *Ocean Development & International Law* 45, no. 2 (2014/04/03 2014): 221–37.

Appiott, Joseph, Amardeep Dhanju, and Biliana Cicin-Sain. "Encouraging Renewable Energy in the Offshore Environment." *Ocean & Coastal Management* 90 (2014/03/01/ 2014): 58–64.

ARENA. "Full Steam Ahead for King Island Wave Power Trial." https://arena.gov.au/blog/king-island-wave-power/.

———. "Uniwave200 King Island Project—Wave Swell." https://arena.gov.au/projects/uniwave200-king-island-project-wave-swell/.

BALANCE. "GIS Tools for Marine Spatial Planning and Management." (2007). https://balance-eu.org/xpdf/balance-interim-report-no-28.pdf.

Becker, Peter, Gavin Burnell, and Tetsuzan Ron. "Using GIS to Improve Coastal Marine Spatial Planning." *Sea Technology* 53 (08/01 2012): 29–35.

Bonar, Paul A. J., Ian G. Bryden, and Alistair G. L. Borthwick. "Social and Ecological Impacts of Marine Energy Development." *Renewable and Sustainable Energy Reviews* 47 (2015/07/01/ 2015): 486–95.

Borthwick, Alistair G. L. "Marine Renewable Energy Seascape." *Engineering* 2, no. 1 (2016/03/01/ 2016): 69–78.

Bureau of Ocean Energy Management. "Comparison of Environmental Effects from Different Offshore Wind Turbine Foundations." (2020). https://www.boem.gov/sites/default/files/documents/environment/Wind-Turbine-Foundations-White%20Paper-Final-White-Paper.pdf.

Center for Wind Energy at James Madison University. "Review of Options for Offshore Foundation Substructures." (2012). https://static1.squarespace.com/static/5b3cf461d274cb109aa488d8/t/5b8b43f370a6adae1d2abc0c/1535853556282/Offshore+Foundations+Substructure+report.pdf.

CNBC. "A Tidal Project in Scottish Waters Just Generated Enough Electricity to Power Nearly 4,000 Homes." https://www.cnbc.com/2020/01/27/tidal-project-generates-electricity-to-power-nearly-4000-homes.html.

Danish Energy Agency. "Danish Experiences from Offshore Wind Development." (2015). https://ens.dk/sites/ens.dk/files/Globalcooperation/offshore_wind_development.pdf.

DEDC. "Puerto Rico Ocean Technology Complex (Protech)." https://www.ddec.pr.gov/en/protech-rfp/.

European Commission. "Energy Sectors and the Implementation of the Maritime Spatial Planning Directive." (2015). https://ec.europa.eu/maritimeaffairs/publications/energy-sectors-and-implementation-maritime-spatial-planning-directive_en.

European Marine Energy Centre. "Tidal Devices." http://www.emec.org.uk/marine-energy/tidal-devices/.

———. "Wave Devices." http://www.emec.org.uk/marine-energy/wave-devices/.

Flynn, Sarah, Will Meaney, Adam M. Leadbetter, Jeffrey P. Fisher, and Caitriona Nic Aonghusa. "Lessons from a Marine Spatial Planning Data Management Process for Ireland." *International Journal of Digital Earth* (2020): 1–19.

Foley, Melissa M., Benjamin S. Halpern, Fiorenza Micheli, Matthew H. Armsby, Margaret R. Caldwell, Caitlin M. Crain, Erin Prahler, et al. "Guiding Ecological Principles for Marine Spatial Planning." *Marine Policy* 34, no. 5 (2010/09/01/ 2010): 955–66.

Fox, Clive J., Steven Benjamins, Elizabeth A. Masden, and Raeanne Miller. "Challenges and Opportunities in Monitoring the Impacts of Tidal-Stream Energy Devices on Marine Vertebrates." *Renewable and Sustainable Energy Reviews* 81 (2018/01/01/ 2018): 1926–38.

Garel, Erwan, Cibran Camba Rey, Óscar Ferreira, and Mark van Koningsveld. "Applicability of the 'Frame of Reference' Approach for Environmental Monitoring of Offshore Renewable Energy Projects." *Journal of Environmental Management* 141 (2014/08/01/ 2014): 16–28.

Gill, Andrew B. "Offshore Renewable Energy: Ecological Implications of Generating Electricity in the Coastal Zone." *Journal of Applied Ecology* 42, no. 4 (2005/08/01 2005): 605–15.

Göteman, Malin, Marianna Giassi, Jens Engström, and Jan Isberg. "Advances and Challenges in Wave Energy Park Optimization—A Review." [In English]. *Frontiers in Energy Research* 8, no. 26 (2020/03/06 2020).

Hammar, Linus, Martin Gullström, Thomas G. Dahlgren, Maria E. Asplund, Ines Braga Goncalves, and Sverker Molander. "Introducing Ocean Energy Industries to a Busy Marine Environment." *Renewable and Sustainable Energy Reviews* 74 (2017/07/01/ 2017): 178–85.

IRENA. "Ocean Thermal Energy Conversion: Technology Brief." (2014). https://www.irena.org/publications/2014/Jun/Ocean-Thermal-Energy-Conversion.

———. "Salinity Gradient Energy Policy Brief." (2014). https://www.irena.org/-/media/Files/IRENA/Agency/Publication/2014/Jun/Salinity_Energy_v4_WEB.pdf

———. "Wind Power Technology Brief." (2016). https://www.irena.org/-/media/Files/IRENA/Agency/Publication/2016/IRENA-ETSAP_Tech_Brief_Wind_Power_E07.pdf

Jones, Lawrence E. *Renewable Energy Integration: Practical Management of Variability, Uncertainty, and Flexibility in Power Grids*. San Diego, CA: Elsevier Science & Technology, 2014.

Maryland Energy Administration. "Offshore Wind Energy in Maryland." https://energy.maryland.gov/Pages/Info/renewable/offshorewind.aspx

Masutani, S. M., and P. K. Takahashi. "Ocean Thermal Energy Conversion (Otec)*." In *Encyclopedia of Ocean Sciences (Second Edition)*, edited by John H. Steele, 167–73. Oxford: Academic Press, 2001.

Murdoch University. "Tidal Barrage & Tidal Turbines." http://www.see.murdoch.edu.au/resources/info/Tech/tidal/

OES-Environmental. "Oes-Environmental 2020 State of the Science Report: Environmental Effects of Marine Renewable Energy Development around the World." (2020). https://tethys.pnnl.gov/publications/state-of-the-science-2020.

Orsted. "Our Experience with Suction Buckets Jacket." https://orsted.com/en/our-business/offshore-wind/wind-technology/suction-bucket-jacket-foundations

Power. "Statkraft Shelves Osmotic Power Project." https://www.powermag.com/statkraft-shelves-osmotic-power-project/

Salvador, Santiago, Luis Gimeno, and F. Javier Sanz Larruga. "The Influence of Maritime Spatial Planning on the Development of Marine Renewable Energies in Portugal and Spain: Legal Challenges and Opportunities." *Energy Policy* 128 (2019/05/01/ 2019): 316–28.

Sánchez, Sergio, José-Santos López-Gutiérrez, Vicente Negro, and M. D. Esteban. "Foundations in Offshore Wind Farms: Evolution, Characteristics and Range of Use. Analysis of Main Dimensional Parameters in Monopile Foundations." *Journal of Marine Science and Engineering* 7, no. 12 (2019).

SIMEC Atlantis Energy. "Meygen." https://simecatlantis.com/projects/meygen/

Simen Bræin, Øystein Skråmestø Sandvik and Stein Erik Skilhagen. "Osmotic Power. From Prototype to Industry—What Will It Take? ." In *3rd International Conference on Ocean Energy*. Bilboa, 2010.

Soukissian, Takvor H., Christos Adamopoulos, Aristides Prospathopoulos, Flora Karathanasi, and Lydia Stergiopoulou. "Marine Renewable Energy Clustering in the Mediterranean Sea: The Case of Pelagos Project." [In English]. *Frontiers in Energy Research* 7, no. 16 (2019/02/28 2019).

State of Green. "Bornholm Can Become the World's First Energy Island." https://stateofgreen.com/en/partners/orsted-a-global-leader-within-green-energy/solutions/bornholm-can-become-the-worlds-first-energy-island/#:~:text=In%20November%202019%2C%20Ørsted%20presented,the%20world's%20first%20energy%20island.

———. "This Is What the World's First Energy Island May Look Like." https://stateofgreen.com/en/partners/state-of-green/news/this-is-what-the-worlds-first-energy-island-may-look-like/.

Sun, Xiaojing, and Diangui Huang. "Wind Power Development in Shanghai, China." *International Journal of Sustainable Energy* 33, no. 1 (2014/01/02 2014): 179–91.

Tidal Lagoon Power. "What Is a Tidal Lagoon?" http://www.tidallagoonpower.com/tidal-technology/what-is-a-tidal-lagoon/#:~:text=A%20tidal%20lagoon%20is%20a,drive%20turbines%20and%20generate%20electricity.

Twomey, Sarah, and Cathal O'Mahony. "Stakeholder Processes in Marine Spatial Planning: Ambitions and Realities from the European Atlantic Experience." In *Maritime Spatial Planning: Past, Present, Future*, edited by Jacek Zaucha and Kira Gee, 295–325. Cham: Springer International Publishing, 2019.

Uihlein, Andreas, and Davide Magagna. "Wave and Tidal Current Energy—A Review of the Current State of Research Beyond Technology." *Renewable and Sustainable Energy Reviews* 58 (2016/05/01/ 2016): 1070–81.

UNESCO. "Marine Spatial Planning: A Step-by-Step Approach Toward Ecosystem-Based Management." (2009). http://www.mspglobal2030.org/wp-content/uploads/2019/03/Marine-spatial-planning-a-step-by-step-approach.pdf.

Young, Michaela. “Building the Blue Economy: The Role of Marine Spatial Planning in Facilitating Offshore Renewable Energy Development.” [In English]. *The International Journal of Marine and Coastal Law* 30, no. 1 (2015/02/17 2015): 148–74.

8

Coastal Water Resources Management

Introduction

Globally, non-point source pollution has contributed to eutrophication in estuarine and coastal waters, resulting in reduced water quality, loss of habitat and natural resources, and hypoxia. Meanwhile, sea-level rise has led to seawater intrusion into surface water and coastal aquifers, diminishing freshwater resources for human use.[1,2] Simultaneously, rising demand for water has led to many locations around the world implementing desalination projects. However, while desalination is secure, reliable, and resilient, the construction and operation of desalination plants can result in various environmental impacts offshore.[3] Furthermore, waterways are transporting significant amounts of plastic pollution into the oceans, degrading marine ecosystems and impacting human health. This chapter will first discuss best management practices (BMPs) to mitigate pollution of surface and groundwater and the ocean before discussing watershed planning to protect water quality. The chapter will then provide an overview of aquifer storage and recovery (ASR). After which, the chapter will discuss a range of mitigation measures desalination projects can adopt to avoid or minimise environmental

R. C. Brears, *Developing the Blue Economy*,
https://doi.org/10.1007/978-3-030-84216-1_8

impacts. Finally, the chapter will discuss the prevention of marine plastic pollution.

Watershed-Level Best Management Practices

A watershed is all the land and water bodies from which water drains to a given point, including a river that reaches the ocean. It is estimated that 80% of pollution to the marine environment comes from the land, with the most significant source being non-point source pollution which occurs because of runoff. Non-point source pollution includes many sources, including farms as well as urban areas and forests. Some water pollution can also start as air pollution, which settles into waterways and oceans. BMPs, as discussed below, describe ways to manage land and activities to mitigate pollution of surface and groundwater and the ocean.[4,5]

Agricultural Best Management Practices

Agricultural BMPs are tools that farmers can use to reduce soil erosion and fertiliser runoff, protect water quality on their farms, and effectively manage animal waste while achieving positive environmental outcomes, including protecting the marine environment. A variety of BMPs commonly used around the world include the following.

Conservation Tillage

Conservation tillage leaves crop residue (plant materials from past harvests) on the soil, reducing the amount of sediment and nutrients that move into the water from agricultural lands. It also reduces wind erosion and dust production. The practice can reduce soil loss by 50% or more as compared to conventional tillage. There are two types of strategies available:

- *Minimum tillage*: Minimum tillage leaves at least 30% of the soil surface covered with plant residue after the tillage or planting operation
- *No-tillage*: This practice leaves the soil undisturbed from harvesting to planting, except for nutrient injection. Crop seeds are planted using a device that opens a trench or slot through the sod or pervious crop residue

Contour Farming

Contour farming aligns all farm tillage, planting, and harvesting practices with the land's contour. The aim is to reduce erosion and surface runoff, which reduces the transport of nutrients and pesticides from the field. Contoured rows retain rainwater, which increases infiltration and reduces runoff.

Cover and Green Manure Crops

Cover crops are crops of close-growing grasses, legumes, or small grains grown primarily for temporary, seasonal soil protection and improvement. Meanwhile, green manure crops are ploughed under and incorporated into the soil to control erosion, add organic matter and nutrients, suppress weeds, and reduce the need for nitrogen fertilisers.

Crop Nutrient Management

Crop nutrient management reduces or prevents nutrient losses from runoff, erosion, and leaching to surface and groundwater resources. It consists of applying nutrients and soil amendments to crops in the:

- *Right amount*: Soil test reports indicate the amount of nutrients that the soil can supply and recommends the amount needed from other sources to produce the indicated crop

- *Right source*: Nitrogen that is structurally part of manure and other organic materials are less prone to short-term loss by leaching because it only becomes available via decomposition
- *Right method*: Nitrogen and phosphorous should be incorporated into the soil before crops are planted/established or banded
- *Right timing*: Nitrogen should be applied frequently in small amounts that are tailored to the plants' immediate needs. Meanwhile, phosphorous is stable when it is mixed in the soil and can be applied when most convenient

Integrated Pest Management

Integrated pest management is an ecologically based strategy of control tactics designed to prevent pest populations from causing negative economic impacts as well as reducing the amount of pesticide used and the amount that moves into the environment. If pests need to be controlled, there are a range of options, including:

- *Biological control*: For example, releasing natural insect enemies
- *Mechanical control*: Ploughing, cultivating
- *Cultural control*: Planting insect-resistant varieties, crop rotation, destroying pest refuge sites
- *Chemical control*: When pesticides must be used, the objective is to select the least toxic product possible and strictly follow its application guidelines on the product label

Vegetative Buffers or Filter Strips

Vegetative buffers or filter strips are strips of grasses or other vegetation placed alongside streams or drainage areas to slow down runoff water, trap sediment, filter nutrients and other pollutants, and promote the infiltration of water into the soil. A filter strip's width depends on the slope and amount of land area delivering water to the strip and the type of vegetation used.

Irrigation Management

Irrigation management promotes the efficient use of irrigation water to produce profitable yield, conserve water, and minimise the leaching of nutrients into groundwater. An irrigation water management plan should use soil-moisture monitoring techniques to determine when irrigation is necessary:

- Applying too much water increases pump costs, reduces water efficiency, and increases the potential for nitrates and pesticides runoff
- Delaying irrigation until plants are water-stressed can reduce yield and make fertilisers and pesticides less effective

Grazing Management

Farmers can adjust grazing intensity, keep livestock out of sensitive areas, and provide alternative sources of water and shade to reduce the impacts of grazing on water quality. There is a wide range of grazing systems that farmers can choose from, including:

- *Rotation*: Intensive grazing followed by resting. Livestock are rotated among two or more pastures during the grazing season
- *Switchback*: Livestock is rotated back and forth between two pastures
- *Rest-rotation*: One pasture is rested for an entire grazing year, or longer with the others grazed on rotation
- *Deferred rotation*: Grazing discontinued on different parts of the farm in succeeding years to allow resting and re-growth
- *Short-duration grazing*: Grazing for 14 days or less with a large herd and high stock density followed by a rest period of 30–90 days
- *High intensity-low frequency*: Heavy, short-duration grazing of all animals on one pasture at a time followed by rotation to another pasture after the forage use goal is met

Animal Feeding Operations Management

Animal feeding operations (AFO) result in accumulated animal wastes, facility wastewater, and storm runoff, all of which can be controlled with proper management techniques. The aim is to minimise the discharge of contaminants in facility wastewater, runoff, and seepage to groundwater. Accumulated animal wastes include manure, litter, or other waste products that can be deposited within the confinement area and are periodically removed by scraping, flushing, or other means to be conveyed to a storage or treatment facility. One of the primary considerations in preventing water pollution from AFOs is the location of the facility. New facilities and expansions of existing ones should take into consideration the following:

- Be located away from surface waters
- Be located away from areas with high leaching potential
- Be located away from critical or sensitive areas
- Be located in areas that minimise odour drift to homes and communities
- Be located away in areas where adequate land is available to regulate runoff movement and increase settling within the facility, for example, vegetated strips, riparian buffers, etc., can reduce delivery of pollutants to surface water by infiltrating, settling, trapping, or transforming nutrients, sediment, and pathogens in runoff leaving the facility

Erosion and Sediment Control

Erosion and sediment control involve constructing an erosion control system to control soil loss from erosion on agricultural land outside of the farmstead or production area. This includes systems utilising terraces, diversions, water and sediment control basins, waterways (grass and lined), and associated earthmoving practices in a system.

Riparian Forest Buffer

A riparian forest buffer is an area of trees and shrubs located adjacent to streams, lakes, ponds, and wetlands. Riparian buffers of sufficient width intercept sediment, nutrients, pesticides, and other surface runoff materials and reduce nutrients and other pollutants in shallow subsurface water flow.

Engaging in Watershed Efforts

Engaging in watershed efforts involves the collaboration of people, stakeholders, and organisations across the watershed to reduce nutrient pollution of water.[6,7,8,9,10,11,12,13,14,15,16]

Case: Louisiana Master Farmer Program
The Louisiana Master Farmer Program is a voluntary certification programme that helps agricultural producers address environmental concerns while enhancing production and resource management skills. The programme helps producers across various agricultural and natural resource enterprises by teaching them more about environmental stewardship, conservation-based production techniques, and sustainability. The programme involves classroom instruction, participation in a field day or workshop, and implementing a comprehensive conservation plan on the entire farming operation. A producer must complete three phases to become a Louisiana Master Farmer, as detailed in Table 8.1.[17]

Table 8.1 Louisiana Master Farmer Program Requirements

Phase	Description
1. Producer attends or views a 6-hour classroom instruction on environmental stewardship issues	Topics covered include: • The Clean Water Act of 1972 • BMPs • Resource conservation planning processes • Spill prevention control and countermeasures
2. Producer attends a conservation-based field day	• Specific BMPs are demonstrated and discussed • It may include pasture walks, soil quality workshops, and other commodity-specific demonstrations
3. Producer requests a farm-specific Resource Management System level conservation plan	• The district conservationist will contact the producer to schedule an initial meeting to discuss the producer's plans, goals and take a resource inventory of the farming operation • The plan is written for the acreage within a subwatershed and only for the acreage that the producer has daily control of

Urban Best Management Practices

In urban settings, the primary goal is to design BMPs that retain runoff at the site and allow water to penetrate the ground where the pollutants can be attached to soil particles and degraded in the vegetation's root zone. A variety of urban BMPs can be implemented to help reduce pollution in waterways, including the following.

Rain Gardens

Rain gardens are small, landscaped depression areas filled with amended native soils or an engineered soil mix planted with trees, shrubs, and other herbaceous vegetation. They are designed to capture and temporarily store stormwater runoff for evaporation, transpiration, and infiltration. This allows sites to reduce stormwater runoff rates and pollutant loads (including total suspended solids, phosphorous, nitrogen, metals, and pathogens). They are typically designed to completely drain within 24 hours of the end of a rainfall event.

Bioretention Areas

Bioretention areas are landscaped shallow depressions that capture and temporarily store stormwater runoff. Runoff is directed into the bioretention area and then filtered through the soil media. They typically consist of a pre-treatment system, surface ponding area, mulch layer, and planting soil media. The depressed area is planted with small- to medium-sized vegetation, including trees, shrubs, and ground cover that can withstand urban environments and tolerate periodic inundation and dry periods. Treatment performance can be enhanced (particularly for nitrogen and other pollutants) by installing deep media with slow infiltration rates.

Water Quality Swales

Water quality swales are vegetated open channels designed to convey runoff without causing erosion while improving stormwater quality. There are two types of swales:

- *Dry swales*: These promote infiltration of runoff and therefore require porous soils
- *Wet swales*: These contain standing water and can use soils with poor drainage or high groundwater conditions

Constructed Stormwater Wetlands

These are engineered shallow-water ecosystems designed to treat stormwater runoff. They are usually implemented in low-lying areas, with them being well-suited to areas along river corridors where water tables are higher. Sediments and nutrients are reduced by sedimentation, chemical and biological conversions, and uptake by wetland plant species. They are designed to maximise the flow path through the wetland to increase residence time and contact with vegetation, soil, and microbes.

Cisterns and Rain Barrels

These are containers that capture rooftop runoff and store water for non-potable uses such as irrigation. Cisterns are large systems that can be self-contained above ground or below ground and are generally larger than 100 gallons and can direct water from one or more downspouts. Rain barrels are smaller systems that direct runoff through a downspout into a barrel that holds less than 100 gallons. Water quality improvements can be made when used with other BMPs, such as bioretention areas.

Permeable Pavement

Permeable pavement is a durable, load-bearing paved surface with small voids or aggregate-filled joints that allow water to drain through to an aggregate reservoir. Stormwater stored in the reservoir layer can then infiltrate underlying soils or drain at a controlled rate via underground drains to other downstream BMPs. Permeable pavement can be developed using modular paving systems (for example, permeable interlocking concrete pavers, concrete grid pavers, or plastic grid systems) or poured in place solutions (for example, pervious concrete or porous asphalt). Permeable pavement systems consistently reduce concentrations and loads of several stormwater pollutants, including heavy metals, oil and grease, sediment, and some nutrients.

Stormwater Planters

These are landscaped planter boxes that are designed to receive stormwater runoff. They consist of planter boxes with waterproof liners, filled with an engineered soil mix, and planted with trees, shrubs, and other herbaceous vegetation. They reduce stormwater runoff rates and pollutant loads.

Stormwater Bump-Outs

Stormwater bump-outs are vegetated kerb extensions that protrude into the street either mid-street or at an intersection, creating a new kerb some distance from the existing kerb. The bump-out is composed of a layer of stone that is topped with soil and plants. An inlet or kerb-cut directs runoff into the bump-out structure, where it can be stored, infiltrated, and taken up by the plants (evapotranspiration).

Stormwater Tree Trenches

Stormwater tree trenches are systems of trees that are connected by an underground infiltration structure. The underground structure is lined with a permeable geotextile fabric, filled with stone or gravel, and topped off with soil and trees. Stormwater flows through a special inlet (storm drain) leading to the stormwater tree trench. The runoff is stored between the empty spaces between the stones, watering the trees and slowly infiltrating through the bottom. If the system's capacity is exceeded, stormwater runoff can bypass it entirely and flow into an existing street inlet.

Green Streets

A green street is designed to integrate a natural system of stormwater management within the public right-of-way. Green streets use various urban BMPs, such as bioretention areas or water quality bioswales, and tree canopies for stormwater interception and water quality improvements. Green streets vary from community to community or even street to street, but they all have the same goal of reducing the amount of stormwater that directly enters waterways.

Green Spaces

Parklands contain significant permeable surfaces that can easily absorb rainwater. If well designed, parks can be enhanced to create hydraulic connections to larger land areas that are mainly impervious, enabling parks to filter stormwater runoff from surrounding roadways and other impervious surfaces.

Urban Forests

An urban forest system includes the trees within an urban area and the ground cover and soil. The system's parts work together to provide significant stormwater volume and pollution control through rainfall interception and intensity reduction, stormwater infiltration and uptake facilitation, and nutrient load reduction. The soil of urban forests filters nutrients and other pollutants from stormwater runoff. Trees need many of the nutrients found in stormwater (nitrogen and phosphorous), and the uptake of these nutrients from the soil by the trees reduces the amount leaching into groundwater.[18,19,20,21,22,23]

Case: Washington DC's Clean Rivers Project
The Clean River Project is the District of Columbia Water and Sewer Authority's (DC Water) ongoing programme to reduce combined sewer overflows (CSOs) into the District waterways. The Clean Rivers Project's BMPs will reduce CSOs by 96% annually throughout the system and reduce the probability of flooding in the areas it serves from 50 to 7% (equivalent to a 15-year storm) in any given year. The project will also reduce nitrogen discharged to the Chesapeake Bay by around one million pounds per annum. DC Water has made its contribution to

reducing CSOs at its facilities by implementing BMPs that are designed to capture precipitation from at least 90% of the storms in an average year, including:

- *Fort Reno Reservoir*: A 42,400 square foot green roof was installed along with 8,400 square feet of pervious pavement
- *Anacostia Water Pumping Station*: 1,000 square feet of pervious pavement was installed along with 1,500 square feet of bioretention
- *East Side Pumping Station*: A 6,600 square foot green roof was installed

DC Water performs regular maintenance of BMPs at its facilities, and pre/post-construction monitoring provides data to improve future BMP design, construction, operation, and maintenance.[24,25]

Forest Best Management Practices

Forestry BMPs minimise water pollution from forestry operations by addressing sediment and sediment transport, which is the primary source of pollution from silviculture. A variety of BMPs that can be implemented include the following.

Preharvest Planning

In the forest environment, building roads, harvesting trees, and preparing sites for new trees disturbs the topsoil, changes the soil's natural filtering action, and modifies the land's topography and drainage features. The result is increased velocity and volume of runoff and increased erosion. A pre-harvest or forest management plan should be developed before any site work. Topographic maps, aerial photographs, and soil surveys should inform the development of the plan. Natural drainage channels, threatened and endangered species habitat, topography, and soil types should be primary considerations when determining the boundaries of timber harvest activities, location and design of roads and landings, selection of harvesting method, and reforestation techniques. Overall, the pre-harvest plan should:

- Identify the area to be harvested
- Locate special areas of protection, such as wetlands and streamside vegetation
- Plan for the proper timing of forestry activities
- Describe management measures for road layout, design, construction, and maintenance
- Describe management measures for harvesting methods and forest regeneration

Streamside Management Zones

Streamside management zones (SMZs) are buffer strips that consist of native vegetation along stream corridors. They filter sediment from silvicultural operations, sustain wildlife and fish populations, and maintain water quality. Their width is based on various factors, including erosiveness of the soil, steepness of the bank slopes, protection of adjacent wetlands, and sensitivity of the fish and wildlife habitat and other critical areas. Disturbances to SMZs should be limited by following the practices of:

- Not constructing roads in SMZs except at designated stream or wetland crossings
- Only operating vehicles on the roads
- Not handling, storing, applying, or disposing of hazardous chemicals, fertilisers, or pesticides in SMZs
- Not depositing waste timber or slash in SMZs

Forest Wetland Protection

Forest wetlands provide many beneficial functions such as sediment trapping, nutrient retention and removal, and groundwater recharge. Normal, ongoing forestry activities, including harvesting, road design and construction, site preparation and regeneration, and chemical management, must be planned and operated to protect forest wetlands' beneficial functions. BMPs to protect forest wetlands include:

- Extending SMZs to incorporate nearby wetlands
- Suspending or limiting operations when soils become saturated
- Maintaining the natural contour of the site and taking action to ensure forestry activities do not immediately or gradually convert the wetland to dry land
- Providing cross-drainage for constructed roads to maintain the natural surface and subsurface flow
- Constructing road fills only when necessary with gravel or crushed rock used as fill to provide for water movement

Road Construction and Maintenance

Well-located and maintained forest roads can minimise the significant source of water pollution associated with silvicultural activities. Poorly designed roads can increase sediment load in streams, increase landslide debris flows, degrade aquatic habitats, and alter and concentrate surface runoff.

- Some BMPs to consider in the layout of the road system are:
  - Plan harvesting activities to minimise the total length of roads required
  - Use existing roads where possible
  - Minimise the changes in the natural stability of the land
  - Where possible, roads should follow the natural contour of the land to avoid extensive cut and fill
  - Minimise the number of stream and wetland crossings
  - Keep the road gradient as low as possible to reduce the velocity of runoff
  - Select the most appropriate road surfacing material to minimise erosion and reduce maintenance costs
- Some BMPs to consider in the construction of the road system are:

  - Whenever possible, construct roads during dry periods or when the ground is frozen. Minimise work during heavy rains and/or wet periods
  - Keep slope stabilisation and erosion and sediment control work as current as possible, including installing drainage features as part of the construction process
  - Place debris, overburden, and other waste materials associated with construction at locations away from streams

- Some BMPs to consider in ensuring proper drainage from road surfaces include:

  - Using, where possible, outsloped roads to drain water directly onto the forest floor
  - If roadside drainage ditches are required, drain water from them to avoid flow concentrations
  - Direct ditch waters onto undisturbed forest floor to allow water to infiltrate and sediment to settle out

Timber Harvesting

Timber harvesting consists of felling the tree, transporting (yarding) to a central accumulation point (landing), and transporting logs offsite along main haul roads. Some BMPs to minimise impacts on water quality include:

- *Implementing skyline systems*: This involves moving logs to the landings by aerial cables in places with long, steep slopes or where it is necessary to protect soils from excessive disturbance
- *Helicopters to move logs to the landings*: This method is used mainly in rugged terrain where good landing sites and roads are not close to the harvest area or where there are sensitive environmental features that may be affected by other yarding methods or new road construction

Revegetation and Forest Regeneration

After an area is harvested, tree stumps, trees too small to harvest, and woody debris are left behind. The area creates a potential fire hazard and leads to increased runoff and stream sedimentation. There are also numerous roads, drainage structures, etc., that are no longer needed, which will become a source of sediment contamination if not properly closed or maintained. A variety of BMPs can be implemented to revegetate and reforest these areas to minimise adverse impacts from past harvesting, including:

- Establishing a vegetative cover planting on erodible areas
- Using native grasses or other plant species to reseed bare-erodible areas (do not introduce invasive, non-native plants)
- Remove unneeded logging roads immediately
- Smooth, grade, and revegetate landings and, where appropriate, main haul roads
- Remove temporary drainage structures and clean permanent drainage structures

Fire Management

Prescribed burning reduces slash, competition for nutrients among seedlings, and fuel for wildfires. When tree species are ecologically dependent on fire for regeneration, fire also serves as a forest management tool. Nonetheless, the intensity and severity of burning and the proportion of watershed burned influences water quality. Specifically, fires that burn intensely on steep slopes close to streams and remove most of the forest floor are most likely to affect water quality by increasing sediment and nutrient pollution. Prescribed burning can avoid significant effects on water quality if appropriate BMPs are utilised, including the following:

- Plan burning to consider the weather, time of year, and fuel conditions
- Do not conduct intense prescribed burns in SMZs and avoid construction of fire lines in SMZs

- Prescribed burns should be only as intense as necessary to achieve the desired objectives

Forest Chemical Management

Many chemical compounds, including pesticides, fertilisers, and fire retardants, are used in forestry, potentially polluting soil and water. A variety of BMPs to ensure the safe use of chemicals includes:

- Ensuring the transport, handling, storage, application, and disposal of pesticides, fertilisers, and fire retardants comply with applicable local, state, national regulations
- Monitoring weather conditions (rain, wind speed, temperature, and humidity) during the application to prevent drift, volatilisation, and surface water runoff
- Ensuring chemicals are not applied in SMZs or wetlands
- Ensuring there is a spill contingency plan which identifies all actions to be taken in the event of a chemical spill[26,27,28,29]

Case: Montana's Forestry Best Management Practices Notification Law

Montana's Forestry Best Management Practices Notification Law became part of the Protection of Forest Resources Law in 1989. The amendments require landowners or operators to notify the Department of Natural Resources and Conservation (DNRC) before conducting forest practices on private lands. The DNRC is charged with providing these landowners and operators with information on BMPs through an on-site visit. Notification is served when an application for a Hazard Reduction Agreement is submitted. The DNRC is also charged with monitoring the application and effectiveness of the BMPs. BMPs have proven to be an effective tool in reducing non-point source pollution from forest harvesting activities. The Forestry Practices Program conducts a biennial audit of the application and effectiveness of BMPS on selected high-risk sites. The first audit in 1990 revealed that 78% of practices met or exceeded BMP standards.

In 1998, the audit results achieved a 96% rating and have been met or exceeded ever since.[30]

Watershed Management Plan

A watershed management plan is a strategy and a work plan for achieving water resource goals in a geographically defined watershed. It provides assessment and management information, including analyses, actions, participants, and resources related to the plan's development and implementation. The main components in a watershed management plan include public concerns, watershed inventory problem identification (including water quality, physical, and social data), identifying sources of problems, selecting critical areas, goals, and objectives, and measuring success. A watershed management plan's primary purpose is to guide watershed coordinators, resource managers, policymakers, and community organisations to restore and protect the quality of lakes, rivers, streams, and wetlands in a given watershed. The plan is intended to be practical with specific recommendations on practices to improve and sustain water quality. The plan also contains "living documents", meaning the plan is re-examined and revised to reflect goals that have been achieved or not. Watershed management plans can be developed to protect coastal and marine ecosystems from various sources of pollution.

Watershed Monitoring

Monitoring is the periodic or continuous collection of data using consistent methods. Water quality monitoring is commonly defined as sampling and analysing water (lake, stream, river, estuary, or ocean) and waterbody conditions. Water quality monitoring can evaluate a water body's physical, chemical, and biological characteristics in relation to human health, ecological conditions, and designated water use. Watershed monitoring is more comprehensive in data collection in that it incorporates water quality and watershed conditions. For example, water quality monitoring conducted on the watershed scale would include monitoring the physical, chemical, and/or biological condition of the waterbody and specific watershed characteristics, such as watershed land use/land cover patterns, that may be related to observed water quality. Watershed monitoring, therefore, evaluates the water resource condition while also providing valuable watershed information to help establish cause-and-effect relationships.

Watershed Monitoring Programme Objectives

The most common objectives in a watershed monitoring programme are:

- *Characterising conditions and trends*: Monitoring data collected regularly can determine water quality and its changes over time. Baseline characterisation can be used to quantify water quality and describe ecological characteristics or processes within the water body or watershed
- *Protecting human health*: Monitoring can advise the citizens of a community about hazardous conditions concerning the use of specific waters. Baseline characterisation can be used to quantify water quality, determine if water quality standards were exceeded, and/or describe ecological condition or processes within the water body or watershed
- *Targeting potential water quality problems*: Monitoring can target areas in a watershed that exhibit more significant potential problems than other areas. For example, a subwatershed that is composed primarily of agricultural land uses may be expected to have higher nutrient loads than a neighbouring forested watershed
- *Designing pollution prevention programmes*: Monitoring can be crucial in designing and implementing a pollution prevention or remediation programme. A pollution prevention or remediation programme can support a risk assessment where monitoring data are used to identify sources and types of pollution in a watershed and evaluate the likelihood of adverse effects. Monitoring information can also be used to select BMPs, to prioritise efforts, and track remedial actions before, during, and after their execution

- *Assessing programme goals*: Monitoring can be used to assess many different programme goals. For example, compliance monitoring is required for wastewater treatment plants and industrial dischargers as part of performance evaluations. Water released from these facilities may be sampled for solids, oxygen demanding wastes, faecal coliform bacteria, metals, nutrients, or other pollutants to determine if levels exceed a permit limitation and pose a threat to public or ecological health
- *Responding to emergencies*: Monitoring can assess the effects of environmental catastrophes such as spills, floods, or droughts. Data may be required to give adequate definition to the water quality problem and the magnitude of the impacts
- *Establishing early warning systems*: The ability to provide early warnings is essential for entire watersheds which may be affected by pollution events. Early warning systems covering accidents and other emergencies should be established for the whole watershed wherever the use of water, potentially threatened by accidental pollution, can be safeguarded through emergency measures. A watershed early warning system can have four main elements:

  - Accident emergency warning systems
  - Hazard identification through databases
  - Models to be used during emergencies
  - Local screening of river water[31,32]

Case: Inshore Water Quality Monitoring in the Great Barrier Reef
Long-term water quality monitoring is a fundamental aspect of determining the status and trends of coastal water quality in the Great Barrier Reef (GBR) lagoon and has been ongoing since 2005. The monitoring helps improve the scientific understanding of how the area is affected by pressures (for example, cyclones, floods, rising ocean temperatures, and land-based runoff) and informs management decisions. Water quality monitoring occurs at inshore sites of four Natural Resource Management Regions: Cape York, Wet Tropics, Burdekin, and Mackay Whitsunday

regions. Collected water samples are analysed for a range of dissolved and particulate nutrients, total suspended solids, chlorophyll a, temperature, salinity, and Secchi depth (a measure of water clarity). Data-logging sensors are also used to continuously measure temperature, salinity, chlorophyll a, and turbidity on selected reefs and at three river mouths. GBR Water Quality Guidelines help managers and scientists understand if water quality can be detrimental to marine ecosystems. A Water Quality Index has been developed to provide an integrated assessment of how water quality is tracking relative to GBR Water Quality Guidelines and shows trends over time. The Index is based on data collected from 2005 onwards and is calculated annually. Scores are derived from five groups of variables:

1. Total suspended solids concentrations, Secchi depth, and turbidity
2. Chlorophyll a concentrations
3. Particulate nitrogen concentrations
4. Particulate phosphorous concentrations
5. Dissolved inorganic nitrogen concentrations in water samples[33,34]

Stakeholder Engagement

Solving various water quality challenges in watersheds requires stakeholders' commitment and participation throughout the watershed's communities, where stakeholders are defined as those who will make decisions, those who will be affected by them, and those who can stop the process if they disagree. Stakeholder engagement is more than just holding public hearings or seeking comments on new regulations. Instead, it involves identifying public concerns and values, developing consensus among affected parties, and producing efficient and effective solutions through an open, inclusive process. The aim is to restore and maintain healthy environmental conditions through community support and cooperative action. A framework for stakeholder engagement should be developed to determine:

- The structure of the group, for example, fully empowered management entity, advisory group, subset of the management committee, or ad hoc group
- How "quiet" or "loud" the stakeholder process needs to be
- How decisions are made, for example, majority vote, consensus, or input received but decisions made by the responsible party
- What the membership of the group will be, for example, one representative from each locality or interest group, a cross-section of the watershed residents, or open to all interested persons
- The roles and responsibilities of the stakeholders, for example, provide input into the scope of efforts, outreach, select management options, represent large constituencies or review and comment on reports, etc.

Education and Outreach

Education and outreach are essential aspects of both watershed planning and encouraging citizen involvements. Increasing the understanding of people's role in affecting water quality can foster skills and motivation to maintain or improve that quality. For any education to be effective, it must be well structured and planned. The first step in starting an education or outreach programme is identifying the outreach goals. The next step is determining the message that people need to understand. Some key questions include:

- Who is the audience for the information, or who needs it the most?
- What does the audience already know?
- What do you want them to know?
- How can you get the audience motivated to care about the message?
- Where or how should the message be delivered?

Overall, various education and outreach programmes and resources can be developed, as listed in Table 8.2.[35,36,37,38]

Table 8.2 Watershed Education and Outreach Programmes

Programme or resource	Description
School curricula	School curricula can be designed around watershed and non-point source issues
Public speakers	Speakers can present at schools, workshops, conferences, and meetings
Public displays	These can be tailored to the audience at a local event or school
Promotional flyers and press releases	Brochures and posters can be delivered through many organisations or created with a special message. They are an easy way to deliver a quick message or to advertise an event. Press releases can be made about a project or event, providing a cost-effective way to get information to a broader audience
Newsletters	Organisation's newsletters may accept outside articles on projects or information
Internet and social media	They provide timely information or refer people to sources of more information
Workshops, forums, conferences, and meetings	These provide continuing education on available resources, updates, data, issues, and thought-provoking ideas
Promotional items	Items with an environmentally friendly tone can get a message across while setting an example

Case: San Francisco's Water Pollution Prevention Programmes

The San Francisco Public Utilities Commission (SFPUC) has developed a range of pollution prevention programmes that employ public education, outreach, and legislation strategies to reduce the amount of pollutants that can enter the bay and ocean via the sewer system:

- *2021 Pollution Prevention Calendar*: The free annual pollution prevention calendar features vivid photos of San Francisco's flora and

landscape while providing everyday tips and resources for preventing pollution at home and in the environment
- *Only Rain Down the Drain*: SFPUC provides a range of tips to prevent pollution of the storm drains and protect the bay and ocean, including tips on:
  - How to prevent stormwater pollution
  - How to use less-toxic gardening products
  - How to properly dispose of pesticides and household chemicals
  - How to wash a car in an environmentally friendly way
- *Toilets are not trashcans!*: The programme educates the public on the top five things that should not be flushed down the toilet[39]

Aquifer Storage and Recovery

ASR is used worldwide to provide seasonal water storage, reduce groundwater overdraft, replenish depleted aquifers, and improve the drinking water supply quality. In coastal areas, ASR systems can act as a hydraulic barrier against saltwater intrusion. There is a wide range of ASR systems, ranging from single well systems to ASR well fields consisting of over 20 wells. The primary method of introducing water into an aquifer is through the following:

- *Infiltration or spreading basins*: Infiltration or spreading basins are the most common form of ASR and have been used worldwide for decades. The water source for spreading basins can range from routine stormwater overland drainage to seasonal high flows from streams and rivers and treated wastewater from municipalities and industries. It involves source water being spread over a land surface and allowed to percolate to the target aquifer. It typically utilises infiltration ponds to enhance the natural percolation of water into the subsurface
- *Injection well*: This is the direct injection of water by a well for subsurface storage and recovery from the same well. Water is injected into an aquifer well during wet periods, during periods of low demand,

or when water quality is good. The injected water displaces the naturally present water in the aquifer, establishing a volume around the well. Water can be recovered immediately after or sometime after infiltration is stopped. Therefore, the periods of infiltration, storage, and recovery can be separate or simultaneous

There are a variety of necessary conditions required for the successful implementation of ASR technology, including:

- The transmissivity (the hydraulic conductivity multiplied by the thickness) of the aquifer should be sufficiently large to allow for the injection and recovery of water at the target well. Low permeable aquitards could reduce the efficiency of an ASR system
- The storage zone of the aquifer should be confined at the top and bottom by less permeable layers so that a large percentage of the injected water can be recovered
- There should be the availability of good quality water for injection to prevent clogging and pollution of native groundwater
- The interactions between the injected water, native water, and soil media should not result in the deterioration of the quality of the stored water
- There needs to be land availability and site accessibility
- There need to be technical capacities on well drilling, maintenance, and operation
- The site needs to be in proximity to electric power infrastructure and water distribution infrastructure

Some of the benefits of ASR include the following:

- The aquifer provides large storage space at no cost
- Aquifer storage does not lose water due to evapotranspiration
- There is a reduced risk of pollution
- Less land is required than for surface reservoirs
- There are fewer impacts on the environment, and impacts are likely to be positive, for example, mitigating the effects of droughts

- Water quality may be further improved with flow passages due to purification characteristics of the subsurface[40,41,42,43,44,45]

> **Case: Aquifer Storage and Recovery in Monterey, California**
> In Monterey, California, ASR entails the diversion of excess Carmel River winter flows, which is then treated and transmitted via the California American Water distribution system to injection/recovery wells in the Seaside Groundwater Basin. Water is diverted from the Carmel River only when it is plentiful and recharges the over-pumped Seaside Basin in wet periods: The overdrafting of groundwater has led to the contamination of aquifers with seawater in the region. Available storage capacity in the Seaside Basin provides an underground reservoir for the diverted water. Water is then pumped out from the Seaside Basin during dry periods to help reduce pumping-related impacts on the Carmel River, including harm to fish and wildlife habitat.[46,47]

Low-Impact Desalination

The purpose of a desalination plant is to collect source seawater in a reliable and sustainable manner to produce desalinated water cost-effectively and with minimal impact on the environment. Desalination is an economical and practical option to meet the rising demand for water, particularly from the commercial and industrial sectors. Nonetheless, while desalination is considered a secure, reliable, and climate-resilient supply of water, the construction and operation of desalination plants can result in various environmental impacts offshore. In response, various mitigation measures can be adopted for avoiding or minimising environmental impacts (Table 8.3). The spatial scales of environmental impacts are defined as:

- *Localised*: Within the area of the project site, within 100 metres of origin

Table 8.3 Potential Offshore Mitigation Measures for Desalination Plants

Phase	Impact and source	Spatial extent	Mitigation measures
Construction	Impacts of construction waste and excess soil from earthworks, construction works	Localised	Recycling of construction wastes and use, for example, excess and leftover soil from digging and piping works, can be used for levelling the site to the desired elevation
	Alteration of the seabed from marine works	Localised	Pipe-jacking for installation of offshore pipelines (intake & outfall) to at least 600 metres from the shoreline; precisely controlled dredging for installation of pipelines from 600 metres; covering of the pipelines and restoration of the original bathymetry
	Sediment resuspension from marine works	Localised to mid-range	Minimal dredging activities; minimisation of drifting and sweeping of dredger suction head by precise positioning control

Phase	Impact and source	Spatial extent	Mitigation measures
	Oil pollution from marine works	Localised to far-range	Prevention of oil pollution from vessels
Operation	Habitat alteration and changes in sediment transport from intake and outfall systems (piping)	Localised	Intake and outfall pipelines laid below the seabed
	Entrainment (passage of small organisms, smaller than the intake screen, with limited or no swimming abilities) and impingement (pinning of large organisms against the intake screens)	Localised to mid-range	Intake heads designed for slow suction velocity
	Debris pollution from intake system (screening system)	Localised	Self-cleaning travelling screen for debris collection; disposal of collected debris in an authorised waste disposal site
	Biological effects of residue chlorine and chlorination by-products from the intake system (biofouling control)	Localised to mid-range	Chlorine dosing (shock treatment) into the intake in the direction of the plant, avoiding chlorine discharge into the sea
	Biological effects of increased seawater salinity from concentrate outfall (reverse osmosis (RO) brine)	Localised to mid-range	Outfall diffuser system designed to enhance initial dilution of the concentrate

(continued)

Table 8.3 (continued)

Phase	Impact and source	Spatial extent	Mitigation measures
	Effects of residual chemicals and particulate matter from concentrate outfall (process streams)	Localised to mid-range	• Minimal use of chemicals (process additives) • Land-based treatment of dual-media filter backwash before discharge and disposal of the sludge in authorised waste disposal sites • Use of environmentally harmless antiscalants • Treatment of limestone reactors washing together with dual-media filter backwash • Use of inorganic solutions only for membrane cleaning and neutralisation of cleaning solutions before discharge

- *Mid-range*: Within the project site and nearby areas, within 1,000 metres of origin
- *Far-range*: Effects beyond project site and nearby areas, beyond 1,000 metres of origin[48,49,50]

Environmental Impact Assessment

An Environmental Impact Assessment (EIA) is a procedure that identifies, describes, evaluates, and develops means of mitigating potential impacts of proposed activities on the environment. An EIA's main objective is to promote environmentally sound and sustainable development by identifying appropriate mitigation measures and alternatives. EIAs should be performed at the earliest stages of project planning and design regarding desalination projects to guide project location, project execution plan, and schedule. Based on the EIA report, the impacts are evaluated. Then a decision can be made for the project according to this evaluation:

- Approve the project according to the set plans if the associated environmental impacts are acceptable according to the authorities' guidelines
- Reconsider the project with given recommendations to mitigate and control the environmental impacts
- Reject the project if the associated environmental impacts are severe[51,52]

Case: Recovering Chemicals from Brine
Researchers at the Massachusetts Institute of Technology have developed a process that can turn concentrated brine into useful chemicals, reducing the amount of brine released into the environment. The approach involves producing sodium hydroxide that can be used to pretreat seawater going into the desalination plant. The sodium hydroxide changes the acidity of the water and helps prevent fouling of the

membranes, which is a significant cause of interruptions and failures in typical RO desalination plants. Already, the desalination industry uses significant volumes of sodium hydroxide. However, cost savings could be achieved if it was made in situ at the plant. The amount required by the plants themselves is far less than the total produced from the brine, so there is potential for it to become a saleable product. Brine can also be used to create hydrochloride acid on-site, with the chemical able to be used for cleaning parts of the desalination plant.[53]

Marine Plastic Pollution

The current approaches to reducing marine plastic pollution are too often focused on individual sectors, making it challenging to address all the root causes of marine plastic pollution. A framework listed in Table 8.4 can be followed for targeting different levels and scales of intervention. At any level, the framework can be implemented at the municipal, regional, national, or transnational level, with a set of guiding questions to prevent marine plastic pollution.[54]

Case: Honolulu Banning Single-use Plastics
In December 2019, the Honolulu City Council passed a single-use plastics ban that went into effect in January 2021 to protect the environment and marine life. From January 1, 2021, food vendors are banned from providing plasticware, including utensils, straws, foam plates, cups, and food containers. In January 2022, additional plastic food ware will be included, and all other businesses will have to adhere to the policy. Failure to comply can result in a fine of up to $1,000 per day. However, exemptions will be made if businesses cannot find suitable non-plastic replacements, such as biodegradable cutlery.[55]

Table 8.4 Framework for Preventing Marine Plastic Pollution

Step	Guiding questions
Step 1. Characterise: Characterise the nature of plastic. The sources, types, behaviour, and impacts of plastic are characterised in this step. In particular, the following questions need to be answered	• How much plastic, of which types, is entering riverine and marine environments? • What are the primary sources of plastics entering riverine and marine environments? • What are the environmental, economic, and social impacts of plastic waste in riverine and marine environments?
Step 2. Engage: Stakeholders are mapped, and then an engagement plan is developed	• Which individuals or groups are affected by plastic pollution and will directly benefit from its prevention? These actors are known as primary stakeholders • Which individuals or groups contribute to plastic pollution, and whose behaviours and practices need to be directly targeted? These actors are known as targeted stakeholders • Which institutions provide or should provide enabling conditions for behavioural changes and benefits to occur and be sustained over time? These actors are known as enabling stakeholders • Are there development partners or financiers whose strategies are aligned with marine plastic pollution prevention? • Are there any other individuals or groups outside of the system who share an interest in marine plastic pollution prevention? These actors are known as external stakeholders

(continued)

Table 8.4 (continued)

Step	Guiding questions
Step 3. Diagnose: The governance systems and behaviours related to marine plastic pollution prevention are analysed	• What are the institutional, legal, and regulatory frameworks, rights, ownership, informal agreements that define the framework for preventing marine plastic pollution at each step of the cycle of plastic production, consumption, and disposal? Are these conflicting or complementary to one another, and where are the gaps in governance that lead to marine plastic pollution? • In addition to the public sector, are there other actors (companies, non-governmental organisations, etc.) that can improve governance related to plastics and what capacity do they have to prevent marine plastic pollution? • Is the behaviour of targeted stakeholders in line with the governance framework (policies, laws, regulations, plans, etc.), or is there a failure of enforcement? • Are there mechanisms for stakeholders to be involved in decision-making? • Are there procedures for conflict resolution?

Step	Guiding questions
Step 4. Design: This step involves defining what needs to change and the interventions required to enact change	• What is the desired long-term impact? • What social, environmental, and/or economic benefits will be reaped by the primary stakeholders? To what extent will pollution be decreased because of intervention? • What changes in practices or behaviours used by the targeted stakeholders are needed to achieve the desired long-term impact? • To what degree are enabling conditions present for the desired changes in practices or behaviours to occur and sustain over time? • What activities and intervention strategies will change the practices or behaviours of the targeted stakeholders and establish the necessary enabling conditions?
Step 5. Act: This involves implementing and funding actions to prevent marine plastic pollution	• What courses of action are needed to establish the conditions and commitments required to ensure the long-term sustainability of marine plastic pollution reduction capacity, funding, and partnerships? • Are there financing partners or mechanisms that will support the implementation of marine plastic pollution reduction actions?

Notes

1. Nancy N. Rabalais et al., "Global Change and Eutrophication of Coastal Waters," *ICES Journal of Marine Science* 66, no. 7 (2009).
2. John E. Hay and Nobuo Mimura, "Sea-Level Rise: Implications for Water Resources Management," *Mitigation and Adaptation Strategies for Global Change* 10, no. 4 (2005).
3. Thomas M. Missimer and Robert G. Maliva, "Environmental Issues in Seawater Reverse Osmosis Desalination: Intakes and Outfalls," *Desalination* 434 (2018).
4. NOAA, "What Is the Biggest Source of Pollution in the Ocean?" https://oceanservice.noaa.gov/facts/pollution.html.
5. USU, "Water Quality Best Management Practices," https://extension.usu.edu/waterquality/protectyourwater/howtoprotectwaterquality/bmps/index.
6. Ibid.
7. Anne Arundel County Soil Conservation District, "Agricultural Best Management Practices," http://www.aascd.org/agriculture/best-management-practices/.
8. ACES, "Agricultural Best Management Practices for Water Quality," https://www.aces.edu/blog/topics/crop-production/agricultural-best-management-practices-for-water-quality/.
9. T. J. Centner et al., "The Adoption of Best Management Practices to Reduce Agricultural Water Contamination," *Limnologica* 29, no. 3 (1999).
10. US EPA, "Protecting Water Quality from Agricultural Runoff," (2015), https://www.epa.gov/sites/production/files/2015-09/documents/ag_runoff_fact_sheet.pdf.
11. NYS Soil and Water Conservation Committee, "Agricultural Best Management Practice Systems Catalogue," (2014), https://www.dec.ny.gov/docs/water_pdf/agriculturebmp.pdf.
12. Anna Lintern et al., "Best Management Practices for Diffuse Nutrient Pollution: Wicked Problems Across Urban and Agricultural Watersheds," *Environmental Science & Technology* 54, no. 15 (2020).

13. NC State Extension, "Best Management Practices for Agricultural Nutrients," https://content.ces.ncsu.edu/best-management-practices-for-agricultural-nutrients.
14. US EPA, "Agricultural Management Practices for Water Quality Protection," (2002), https://cfpub.epa.gov/watertrain/pdf/modules/Agriculture.pdf.
15. R. C. Brears, *Water Resources Management: Innovative and Green Solutions* (De Gruyter, 2021).
16. *Regional Water Security* (Wiley, 2021).
17. LSU AgCenter, "Louisiana Master Farmer Program," https://www.lsuagcenter.com/topics/environment/conservation/master_farmer.
18. Kansas State University, "Urban Water Quality Best Management Practices," (2017), https://bookstore.ksre.ksu.edu/pubs/MF2732.pdf.
19. US EPA, "Coastal Stormwater Management through Green Infrastructure a Handbook for Municipalities," (2014), https://www.epa.gov/sites/production/files/2015-09/documents/massbays_handbook_combined_508-opt_1.pdf.
20. Center for Watershed Protection, "Coastal Stormwater Supplement to the Georgia Stormwater Management Manual," (2009), http://documents.atlantaregional.com/gastormwater/Georgia-CSS-Final-Apr-09.pdf.
21. US EPA, "Stormwater to Street Trees: Engineering Urban Forests for Stormwater Management," (2013), https://www.epa.gov/sites/production/files/2015-11/documents/stormwater2streettrees.pdf#page27.
22. USDA, "Urban Forest Systems and Green Stormwater Infrastructure," (2020), https://www.fs.usda.gov/sites/default/files/fs_media/fs_document/Urban-Forest-Systems-GSI-FS-1146.pdf#page27.
23. R. C. Brears, *Blue and Green Cities: The Role of Blue-Green Infrastructure in Managing Urban Water Resources* (Palgrave Macmillan UK, 2018).
24. DC Water, "Clean Rivers Project," https://www.dcwater.com/cleanrivers.

25. "Green Infrastructure Projects at DC Water Facilities," https://www.dcwater.com/projects/green-infrastructure-projects-dc-water-facilities.
26. US EPA, "Forestry Best Management Practices in Watersheds," https://cfpub.epa.gov/watertrain/moduleFrame.cfm?module_id=34&parent_object_id=1517&object_id=1517.
27. USDA, "National Best Management Practices for Water Quality Management on National Forest System Lands," (2012), https://www.fs.fed.us/naturalresources/watershed/pubs/FS_National_Core_BMPs_April2012.pdf.
28. Brooke M. Warrington et al., "Forestry Best Management Practices Relationships with Aquatic and Riparian Fauna: A Review," *Forests* 8, no. 9 (2017).
29. UNH, "Best Management Practices for Forestry: Protecting New Hampshire's Water Quality," (2005), https://extension.unh.edu/resources/representation/Resource000248_Rep267.pdf.
30. Montana Department of Natural Resources and Conservation, "Best Management Practices—BMP," http://dnrc.mt.gov/divisions/forestry/forestry-assistance/forest-practices/best-management-practices-bmp-2.
31. US EPA, "Overview of Watershed Monitoring," (2007), https://cfpub.epa.gov/watertrain/pdf/modules/monitoring.pdf.
32. WMO, "Planning of Water Quality Monitoring Systems," (2013), https://www.wmo.int/pages/prog/hwrp/publications/Technical_report_series/TR-No3water_quality_monitoring_systems.pdf.
33. eAtlas, "Marine Monitoring Program—Inshore Water Quality Monitoring (Aims)," https://eatlas.org.au/rrmmp/gbr-aims-inshore-water-quality.
34. Australian Goverment Great Barrier Reef Marine Park Authority, "Marine Monitoring Program," https://www.gbrmpa.gov.au/our-work/our-programs-and-projects/marine-monitoring-program.
35. New Hampshire Department of Environmental Services, "Non-point Source Pollution: A Guide for Citizens and Town Officials," (2004), https://www.allenstownnh.gov/sites/g/files/vyhlif241/f/file/file/nh_bmp_guide.pdf.

36. US EPA, "Getting in Step: Engaging Stakeholders in Your Watershed," (2013), https://cfpub.epa.gov/npstbx/files/stakeholderguide.pdf%20.
37. Indiana Department of Environmental Management, "Watershed Management Plans," https://www.in.gov/idem/nps/3180.htm#:~:text=A%20watershed%20plan%20is%20a,and%20implementation%20of%20the%20plan.
38. R. C. Brears, *Urban Water Security* (Chichester, UK; Hoboken, NJ: Wiley, 2016).
39. SFPUC, "Water Pollution Prevention," https://sfwater.org/index.aspx?page=96.
40. UNESCO-IHE Deltares, and Jahangirnagar University, "Swibangla—Managing Salt Water Intrusion Impacts in Bangladesh," (2015), https://www.deltares.nl/app/uploads/2015/04/1207671-000-BGS-0016-r-SWIBANGLA-def.pdf.
41. Chunhui Lu et al., "Recovery Efficiency of Aquifer Storage and Recovery (ASR) with Mass Transfer Limitation," *Water Resources Research* 47, no. 8 (2011).
42. PREPARED, "Aquifer Storage and Recovery (ASR). Design and Operational Experiences for Water Storage Through Wells," (2013), https://www.researchgate.net/publication/295854388_Aquifer_Storage_and_Recovery_ASR_Design_and_operational_experiences_for_water_storage_through_wells.
43. US EPA, "Aquifer Recharge and Aquifer Storage and Recovery," https://www.epa.gov/uic/aquifer-recharge-and-aquifer-storage-and-recovery.
44. DEMEAU, "Managing Aquifer Recharge," (2015), https://demeau-fp7.eu/sites/files/DEMEAU_MAR_brochure.pdf.
45. IDB, "Aquifer Storage and Recovery: Improving Water Supply Security in the Caribbean Opportunities and Challenges," (2019), https://publications.iadb.org/publications/english/document/Aquifer_Storage_and_Recovery_Improving_Water_Supply_Security_in_the_Caribbean_Opportunities_and_Challenges_en.pdf.
46. Monterey Peninsula Water Management District, "Aquifer Storage & Recovery," https://www.mpwmd.net/water-supply/aquifer-storage-recovery/.

47. DeepWater Desal, "Regional Water Challenges," https://www.deepwaterdesal.com/regional-water-challenges.htm.
48. Water Reuse Association, "Overview of Desalination Plant Intake Alternatives," (2011), https://watereuse.org/wp-content/uploads/2015/10/Intake_White_Paper.pdf.
49. F. Lokiec, "Sustainable Desalination: Environmental Approaches," in *The International Desalination Association World Congress on Desalination and Water Reuse* (Tianjin, China2013).
50. Esmaeil Ahmadi et al., "The Role of Renewable Energy Resources in Sustainability of Water Desalination as a Potential Fresh-Water Source: An Updated Review," *Sustainability* 12, no. 13 (2020).
51. Khaled Elsaid et al., "Environmental Impact of Desalination Processes: Mitigation and Control Strategies," *Science of The Total Environment* 740 (2020).
52. Ahmadi et al., "The Role of Renewable Energy Resources in Sustainability of Water Desalination as a Potential Fresh-Water Source: An Updated Review."
53. MIT, "Turning Desalination Waste into a Useful Resource," https://news.mit.edu/2019/brine-desalianation-waste-sodium-hydroxide-0213.
54. SIWI, "Source-to-Sea Framework for Marine Litter Prevention: Preventing Plastic Leakage from River Basins," (2019), https://www.siwi.org/wp-content/uploads/2019/08/Marine-Litter-Prevention_2019_WEBB.pdf.
55. New York City Food Policy Center, "Honolulu Bans Single-Use Plastics Starting January 2021," https://www.nycfoodpolicy.org/food-policy-center-honolulu-plastics-ban/.

References

ACES. "Agricultural Best Management Practices for Water Quality." https://www.aces.edu/blog/topics/crop-production/agricultural-best-management-practices-for-water-quality/.

Ahmadi, Esmaeil, Benjamin McLellan, Behnam Mohammadi-Ivatloo, and Tetsuo Tezuka. "The Role of Renewable Energy Resources in Sustainability of Water Desalination as a Potential Freshwater Source: An Updated Review." *Sustainability* 12, no. 13 (2020).

Anne Arundel County Soil Conservation District. "Agricultural Best Management Practices." http://www.aascd.org/agriculture/best-management-practices/.

Australian Goverment Great Barrier Reef Marine Park Authority. "Marine Monitoring Program." https://www.gbrmpa.gov.au/our-work/our-programs-and-projects/marine-monitoring-program.

Brears, R. C. *Blue and Green Cities: The Role of Blue-Green Infrastructure in Managing Urban Water Resources.* Palgrave Macmillan UK, 2018.

———. *Regional Water Security.* Wiley, 2021.

———. *Urban Water Security.* Chichester, UK; Hoboken, NJ: Wiley, 2016.

———. *Water Resources Management: Innovative and Green Solutions.* De Gruyter, 2021.

Center for Watershed Protection. "Coastal Stormwater Supplement to the Georgia Stormwater Management Manual." (2009). http://documents.atlantaregional.com/gastormwater/Georgia-CSS-Final-Apr-09.pdf.

Centner, T. J., J. E. Houston, A. G. Keeler, and C. Fuchs. "The Adoption of Best Management Practices to Reduce Agricultural Water Contamination." *Limnologica* 29, no. 3 (1999/07/01/ 1999): 366–73.

DC Water. "Clean Rivers Project." https://www.dcwater.com/cleanrivers.

———. "Green Infrastructure Projects at DC Water Facilities." https://www.dcwater.com/projects/green-infrastructure-projects-dc-water-facilities.

DeepWater Desal. "Regional Water Challenges." https://www.deepwaterdesal.com/regional-water-challenges.htm.

Deltares, UNESCO-IHE, and Jahangirnagar University. "Swibangla—Managing Salt Water Intrusion Impacts in Bangladesh." (2015). https://www.deltares.nl/app/uploads/2015/04/1207671-000-BGS-0016-r-SWIBANGLA-def.pdf.

DEMEAU. "Managing Aquifer Recharge." (2015). https://demeau-fp7.eu/sites/files/DEMEAU_MAR_brochure.pdf.

eAtlas. "Marine Monitoring Program—Inshore Water Quality Monitoring (Aims)." https://eatlas.org.au/rrmmp/gbr-aims-inshore-water-quality.

Elsaid, Khaled, Enas Taha Sayed, Mohammad Ali Abdelkareem, Ahmad Baroutaji, and A. G. Olabi. "Environmental Impact of Desalination Processes: Mitigation and Control Strategies." *Science of The Total Environment* 740 (2020/10/20/ 2020): 140125.

EPA, US. "Forestry Best Management Practices in Watersheds." https://cfpub.epa.gov/watertrain/moduleFrame.cfm?module_id=34&parent_object_id=1517&object_id=1517.

Hay, John E., and Nobuo Mimura. "Sea-Level Rise: Implications for Water Resources Management." *Mitigation and Adaptation Strategies for Global Change* 10, no. 4 (2005/10/01 2005): 717–37.

IDB. "Aquifer Storage and Recovery: Improving Water Supply Security in the Caribbean Opportunities and Challenges." (2019). https://publications.iadb.org/publications/english/document/Aquifer_Storage_and_Recovery_Improving_Water_Supply_Security_in_the_Caribbean_Opportunities_and_Challenges_en.pdf.

Indiana Department of Environmental Management. "Watershed Management Plans." https://www.in.gov/idem/nps/3180.htm#:~:text=A%20watershed%20plan%20is%20a,and%20implementation%20of%20the%20plan.

Kansas State University. "Urban Water Quality Best Management Practices." (2017). https://bookstore.ksre.ksu.edu/pubs/MF2732.pdf.

Lintern, Anna, Lauren McPhillips, Brandon Winfrey, Jonathan Duncan, and Caitlin Grady. "Best Management Practices for Diffuse Nutrient Pollution: Wicked Problems Across Urban and Agricultural Watersheds." *Environmental Science & Technology* 54, no. 15 (2020/08/04 2020): 9159–74.

Lokiec, F. "Sustainable Desalination: Environmental Approaches." In *The International Desalination Association World Congress on Desalination and Water Reuse*. Tianjin, China, 2013.

LSU AgCenter. "Louisiana Master Farmer Program." https://www.lsuagcenter.com/topics/environment/conservation/master_farmer.

Lu, Chunhui, Pengfei Du, Yiming Chen, and Jian Luo. "Recovery Efficiency of Aquifer Storage and Recovery (ASR) with Mass Transfer Limitation." *Water Resources Research* 47, no. 8 (2011/08/01 2011).

Missimer, Thomas M., and Robert G. Maliva. "Environmental Issues in Seawater Reverse Osmosis Desalination: Intakes and Outfalls." *Desalination* 434 (2018/05/15/ 2018): 198–215.

MIT. "Turning Desalination Waste into a Useful Resource." https://news.mit.edu/2019/brine-desalianation-waste-sodium-hydroxide-0213.

Montana Department of Natural Resources and Conservation. "Best Management Practices—BMP." http://dnrc.mt.gov/divisions/forestry/forestry-assistance/forest-practices/best-management-practices-bmp-2.

Monterey Peninsula Water Management District. "Aquifer Storage & Recovery." https://www.mpwmd.net/water-supply/aquifer-storage-recovery/.

NC State Extension. "Best Management Practices for Agricultural Nutrients." https://content.ces.ncsu.edu/best-management-practices-for-agricultural-nutrients

New Hampshire Department of Environmental Services. "Nonpoint Source Pollution: A Guide for Citizens and Town Officials." (2004). https://www.allenstownnh.gov/sites/g/files/vyhlif241/f/file/file/nh_bmp_guide.pdf.

New York City Food Policy Center. "Honolulu Bans Single-Use Plastics Starting January 2021." https://www.nycfoodpolicy.org/food-policy-center-honolulu-plastics-ban/

NOAA. "What Is the Biggest Source of Pollution in the Ocean?" https://oceanservice.noaa.gov/facts/pollution.html.

NYS Soil and Water Conservation Committee. "Agricultural Best Management Practice Systems Catalogue." (2014). https://www.dec.ny.gov/docs/water_pdf/agriculturebmp.pdf.

PREPARED. "Aquifer Storage and Recovery (ASR). Design and Operational Experiences for Water Storage Through Wells." (2013). https://www.researchgate.net/publication/295854388_Aquifer_Storage_and_Recovery_ASR_Design_and_operational_experiences_for_water_storage_through_wells.

Rabalais, Nancy N., R. Eugene Turner, Robert J. Díaz, and Dubravko Justić. "Global Change and Eutrophication of Coastal Waters." *ICES Journal of Marine Science* 66, no. 7 (2009): 1528–37.

SFPUC. "Water Pollution Prevention." https://sfwater.org/index.aspx?page=96.

SIWI. "Source-to-Sea Framework for Marine Litter Prevention: Preventing Plastic Leakage from River Basins." (2019). https://www.siwi.org/wp-content/uploads/2019/08/Marine-Litter-Prevention_2019_WEBB.pdf.

UNH. "Best Management Practices for Forestry: Protecting New Hampshire's Water Quality." (2005). https://extension.unh.edu/resources/representation/Resource000248_Rep267.pdf

US EPA. "Agricultural Management Practices for Water Quality Protection." (2002). https://cfpub.epa.gov/watertrain/pdf/modules/Agriculture.pdf.

———. "Aquifer Recharge and Aquifer Storage and Recovery." https://www.epa.gov/uic/aquifer-recharge-and-aquifer-storage-and-recovery.

———. "Coastal Stormwater Management through Green Infrastructure a Handbook for Municipalities." (2014). https://www.epa.gov/sites/production/files/2015-09/documents/massbays_handbook_combined_508-opt_1.pdf.

———. "Getting in Step: Engaging Stakeholders in Your Watershed." (2013). https://cfpub.epa.gov/npstbx/files/stakeholderguide.pdf%20.

———. "Overview of Watershed Monitoring." (2007). https://cfpub.epa.gov/watertrain/pdf/modules/monitoring.pdf.

———. "Protecting Water Quality from Agricultural Runoff." (2015). https://www.epa.gov/sites/production/files/2015-09/documents/ag_runoff_fact_sheet.pdf

———. "Stormwater to Street Trees: Engineering Urban Forests for Stormwater Management." (2013). https://www.epa.gov/sites/production/files/2015-11/documents/stormwater2streettrees.pdf#page27.

USDA. "National Best Management Practices for Water Quality Management on National Forest System Lands." (2012). https://www.fs.fed.us/naturalresources/watershed/pubs/FS_National_Core_BMPs_April2012.pdf.

———. "Urban Forest Systems and Green Stormwater Infrastructure." (2020). https://www.fs.usda.gov/sites/default/files/fs_media/fs_document/Urban-Forest-Systems-GSI-FS-1146.pdf#page27.

USU. "Water Quality Best Management Practices." https://extension.usu.edu/waterquality/protectyourwater/howtoprotectwaterquality/bmps/index.

Warrington, Brooke M., W. M. Aust, Scott M. Barrett, W. M. Ford, C. A. Dolloff, Erik B. Schilling, T. B. Wigley, and M. C. Bolding. "Forestry Best Management Practices Relationships with Aquatic and Riparian Fauna: A Review." *Forests* 8, no. 9 (2017).

Water Reuse Association. "Overview of Desalination Plant Intake Alternatives." (2011). https://watereuse.org/wp-content/uploads/2015/10/Intake_White_Paper.pdf.

WMO. "Planning of Water Quality Monitoring Systems." (2013). https://www.wmo.int/pages/prog/hwrp/publications/Technical_report_series/TR-No3water_quality_monitoring_systems.pdf.

9

Blue Carbon Ecosystems and Ecosystem-Based Adaptation

Introduction

Blue carbon refers to carbon dioxide removed from the atmosphere by coastal and marine ecosystems, including coastal wetlands, mangroves, and seagrasses. Blue carbon ecosystems are being lost at a fast rate due to various trends. When these ecosystems degrade, they contribute to climate change by releasing stored carbon into the atmosphere.[1,2,3,4] Furthermore, coastal ecosystems are exposed to sea-level rise, flooding, erosion, and habitat loss due to climate change and land-use change. When coastal ecosystems are degraded or lost, their natural capacity to buffer the impacts of extreme events is lost, resulting in damage to infrastructure and property along the coastal edge.[5] This chapter will first discuss various measures to conserve and restore blue carbon ecosystems. The chapter will then discuss seaweed production as a climate mitigation solution. Finally, the chapter will discuss the concept of ecosystem-based adaptation (EbA) and specific EbA approaches in coastal and marine areas.

R. C. Brears, *Developing the Blue Economy*,
https://doi.org/10.1007/978-3-030-84216-1_9

Blue Carbon Ecosystems

Blue carbon ecosystems are being lost at a fast rate due to deforestation, increasing coastal population sizes and coastal development, agriculture and aquaculture, sedimentation and siltation, and the impacts of climate change, including rising sea levels and extreme weather events. When these ecosystems degrade, they turn from carbon sinks to contributors to climate change by releasing stored carbon into the atmosphere.[6,7,8,9]

Blue carbon ecosystems' dynamic nature means that sequestration rates in coastal wetlands, mangroves, and seagrasses vary across spatial and temporal scales. Carbon sequestration in coastal wetlands and mangroves rise or fall depending on the local level and extent of tidal inundation, sediment accumulation, salinity, climate variability, and microbe activity. Meanwhile, seagrass sequestration is influenced by hydrological depth, processes, turbidity, and ecological aspects, such as sediment size, structural complexity within meadows, and variable primary productivity rates present in different coastal areas.[10] Table 9.1 provides an estimate of the mean and maximum (in brackets) of the area covered by blue carbon sinks and the annual organic carbon burial rates. Carbon burial rates are presented per hectare (mean, range, and the upper confidence limit of the mean of individual ecosystem estimates, in brackets) and globally (as reported ranges of mean rates of global carbon burial and, in brackets, an upper estimate).[11]

Blue carbon strategies focus on preserving and enhancing the organic carbon stocks and organic carbon burial capacity of wetlands, mangroves, and seagrass, particularly within their soil.[12] Unlike terrestrial soils, the soils of blue carbon ecosystems are primarily anaerobic, resulting in carbon incorporated into the soil, which decomposes very slowly

Table 9.1 Mean and Maximum Estimates of Area Covered by Blue Carbon Sinks

Vegetated habitats	Area (Million km^2)	Organic carbon burial	
		Tonne C ha-1 y-1	Tg C y-1
Mangroves	0.17 (0.3)	1.39, 0.20–6.54 (1.89)	17–23.6 (57)
Salt Marsh	0.4 (0.8)	1.51, 0.18–17.3 (2.37)	60.4–70 (190)
Seagrass	0.33 (1.7)	0.83, 0.56–1.82 (1.37)	27.4–44 (82)

and can be stored for hundreds or even thousands of years. Also, the high salinity in many blue carbon ecosystems limits methane production.[13] The conservation and creation of blue carbon ecosystems also provide numerous benefits and services for climate change adaptation along coasts, including protection from storms, prevention of coastal flooding and shoreline erosion, regulation of water quality, provision of habitat for commercially important fisheries, and food security for coastal communities.[14] Specific measures to conserve and restore blue carbon ecosystems are as follows.

Coastal Wetland Restoration

Coastal wetlands contain partially and fully submerged vegetation suited to both fresh and saltwater. In addition to carbon sequestration, coastal wetlands, when healthy, filter nutrients and sediment from passing water, protect against wave damage and erosion, and diminish flooding by holding excess storm waters.[15,16] Coastal wetland restoration refers to the return of wetland from a disturbed or altered state caused by human activities to pristine status. Methods for coastal wetland restoration can be classified as follows:

- *Active*: This requires humans to control and intervene regularly to restore, recreate, or improve the wetlands' community structure and ecosystem processes. Activities include reshaping the topography of the wetlands, rechannelling the water flow through water control facilities, such as dikes, soil transplantation, and artificially planting vegetation
- *Passive*: This involves the elimination of influencing factors that lead to the degradation or destruction of coastal wetlands and the restoration of coastal wetlands to a healthy state under natural conditions
- *Creation*: This refers to the process of turning lands on sites where no wetlands existed previously into a wetland. It requires appropriate local conditions, particularly suitable hydrology[17,18]

A set of recommendations are suggested to ensure the success of ecological restoration of coastal wetlands:

- *Right species at the right place*: Site selection is critical for successful coastal wetland restoration, with local species prioritised. Most of the species' establishment needs a critical threshold of hydrodynamics, such as flow speed and direction, inundation period and depth, and even suspended sediment content. These factors for different species should be assessed first before taking actions
- *Make use of the ecosystem's self-organisation strength*: Once the vegetation community is established, positive feedback between sediment accumulation and vegetation expansion and succession will be formed. As such, it is more cost-effective to build small patches than to plant seedlings on a large scale
- *Identify reliable economic benefits for the local community*: Ecotourism is one of a few well-known economic activities that can stimulate the protection and restoration of coastal wetlands[19]

Before setting out to restore degraded coastal wetlands, a feasibility analysis should be performed. It is also essential to identify how humans can actively facilitate the recovery process. Furthermore, wetland managers and decision-makers need to consider how to assess whether the restoration project is successful. An indicator system to measure success should be developed with wetland restoration goals identified to ensure that the selected success indicators are reasonable. Furthermore, the indicator system should be able to assess coastal wetland structure (for example, landscape or community composition, disturbance, and evolution), function (i.e., productivity and ecological service functions), and disturbance (for example, reclamation, drainage, and intensity, range, and frequency of disturbance).[20,21]

Mangrove Restoration

Mangroves are found within intertidal areas. They are a type of salt-tolerant vegetation that includes trees and shrubs with extensive below-surface root structures and deep sediments. These deep layers of sediment store anaerobic carbon. Carbon stocks of mangroves are not uniform.

The depth of carbon-rich soil relates to the geomorphology of an environment. There are also differences in carbon storage between estuarine and oceanic mangroves where the substrate material differs. In addition to carbon sequestration, mangroves provide a range of co-benefits to human populations, with coastal communities relying on mangroves' provisioning services, for example, the extraction of construction materials and fuelwood and the capture of food sources such as shellfish. Mangrove restoration follows the same principles as restoring terrestrial forests. It involves collecting propagules, planting them in a prepared nursery, and transplanting the propagules to the desired location. Other mangrove restoration actions include:

- Addressing the causes of degradation and natural impediments to mangrove settlements and growth
- Improving land-use practices to reduce nutrient and sediment runoff
- Hydrological restoration, including restoring tidal hydrology through excavation or backfilling and/or reconnecting blocked areas to normal tide influences
- Restoring hydrological regimes and sediment flow to help mangroves keep up with the sea-level rise and support carbon sequestration

However, mangrove restoration initiatives often fail due to inadequate site assessment, lack of consideration of the site quality, and a lack of long-term monitoring of mangrove restoration projects. The failures can be categorised in several ways, in particular:

- Extreme changes in site conditions, for example, soil factors, particularly salinity, hydrology, and sedimentation
- Inappropriate restoration techniques, for instance, species-site mismatch, poor quality propagules, poor nursery establishment and management, poor site preparation, inappropriate transplants
- Failure to involve all stakeholders, especially local communities, and relevant government institutions

A set of guidelines for ensuring successful mangrove restoration initiatives are as follows:

1. *Project planning*: Restoration campaigns should be initiated by starting with the identification of the problem. There should also be consensus with stakeholders on the definition and formulation of the purpose of implementing a restoration project and the agreed-upon objective(s). Furthermore, stakeholder engagement should be extended across the whole restoration process to ensure that the interests of local communities are recognised, appreciated, and safeguarded
2. *Project implementation*: There needs to be an understanding of why natural regeneration is not occurring or sufficient and then make adjustments to the site or find solutions to social issues. Also, planting should be close to where species naturally occur, local communities should be involved in the planting, the site should be protected from people and animals, and the site should be monitored over the long term (usually five years) and adjustments made
3. *Project monitoring and evaluation*: In most restoration projects, the perception is that once mangroves are planted, they will grow without further monitoring. This is likely to fail because mangroves are dynamic ecosystems. Therefore, a range of issues should be considered during the monitoring and evaluation phase:

    a. Revisit the restoration site to assess the performance
    b. Identify who will implement the monitoring plan
    c. Assess secondary succession (flora and fauna)
    d. Evaluate the cost of restoration projects[22,23,24,25]

Seagrass Restoration

Seagrasses are fully submerged flowering plants that can grow in meadows and are found in nearshore coastal areas. In addition to carbon sequestration, seagrasses provide various ecosystem services, including habitat for many fish and invertebrate species and water filtration by holding nutrients and sediment in their grassy biomass.[26,27] Overall, they are highly productive ecosystems and play a vital role as habitats supporting high biodiversity.[28,29,30] Seagrass restoration can be carried out using seagrass seedlings germinated in vitro from seeds, the optimisation of which can provide large quantities of seedlings.[31] Projects that plant seeds use various methods, including spreading by hand at the water surface, planting seeds in coconut matting, and planting seeds and sediment in hessian bags anchored on the seabed.[32] Projects can also transplant seagrass from donor sites to transplant sites, with extensive initial transplants increasing the likelihood of success.[33] Meanwhile, restoration efforts to recover seagrass meadows include removing impacts and improving habitat conditions to allow natural recovery.[34] A set of guidelines are recommended for seagrass conservation and restoration projects:

1. *Reverse habitat degradation*: Before any restoration or restorative effort, the causes of degradation should be known and alleviated or reversed
2. *Select appropriate habitat*: The transplantation location should preferably have a history of seagrass growth, its depth should be similar to nearby natural seagrass beds, and the habitat requirements should be met as much as possible
3. *Select an appropriate donor population*: The plants should be adapted to the local environmental conditions, and the transplantation should have sufficient genetic variation to be able to adapt to environmental changes and avoid interbreeding
4. *Spread risks*: In dynamic coastal and estuarine environments, spreading the risk of plant losses in time and space is essential. Natural populations survive extreme conditions (storms, droughts, salinity fluctuations, etc.) by maintaining genetic variation and implementing multiple reproductive or growth strategies and so on. The spreading of risks can be done at:

   a. A kilometre scale, for example, by transplanting to areas that differ in hydrodynamic exposure. These sites will probably differ in habitat characteristics
   b. A local scale, for example, by applying replicates at the distance of tens of metres and hundreds of metres, but also by planting at different tidal depths
   c. A temporal scale by transplanting in different years, for example, by transplanting to the same site over two or more years

d. A temporal scale by transplanting at different dates, for example, by transplanting in different months. Usually, only a few months per year yield optimal transplantation results

5. *Account for hydrodynamics*: To cope with hydrodynamic stress and disturbances, the habitat can potentially be locally stabilised with, for example, miniature dikes around small plots and creating shell armouring of the sediment[35,36,37,38]

Case: International Blue Carbon Initiative Mitigating Climate Change

The International Blue Carbon Initiative is a coordinated, global programme that focuses on mitigating climate change by conserving and restoring coastal and marine ecosystems. The initiative focuses on mangroves, salt marshes, and seagrasses with strategically designed and implemented field projects demonstrating the viability of blue carbon and facilitating the development of practical, science-based methodologies to build local and national capacity to protect and manage coastal ecosystems. Projects include the following:

- *Kaimana Coastal Conservation and Development, Indonesia*: This project is a collaboration between Conservation International, University of Queensland, University of Texas Pan-American, The State University of Papua, and the Ministry of Marine Affairs and Fisheries. The project serves as a pilot to demonstrate the climate mitigation potential and viability of blue carbon projects to international policy, national governments, and local managers
- *Tomago Wetland Restoration, Australia*: This project, located in the Hunter River Estuary, a Ramsar-listed wetland near the City of Newcastle, is a collaboration between NSW Fisheries, the National Parks and Wildlife Service, the Hunter Regional Local Land Services, and the Water Research Laboratory at the University of New South Wales. It involves the reintroduction of tidal waters to impounded wetlands to facilitate fish passage and the restoration of habitat for shorebirds[39]

Seaweed Production

There is potential for seaweed farming to serve as a global climate mitigation solution and may be a viable option for forward-looking countries or economies that set ambitious carbon emission reduction targets while supporting blue economic development.[40] In particular, seaweed can be a feedstock for bioenergy applications, with current global marine agronomy producing around 30 million metric dry tonnes of seaweed per annum. Seaweed can also help reduce the emissions from agriculture by substituting synthetic fertiliser and lowering methane emissions from cattle when included in cattle feed.[41,42]

Seaweeds are generally cultivated in sheltered bays, with seaweeds obtaining nutrients directly from the seawater. Therefore, it is crucial to have currents that flush the site in which the seaweeds are placed. Small cuttings are attached to a line with short lengths of string in shallow waters, with the seaweed harvested during the two spring tide periods each month. In deeper waters, longlines are used with lines reaching lengths of 50 metres. The growth rate of seaweed in suitable habitat is significant: a single cutting of 100 grams can reach a weight of one kilogram in 20–40 days, yielding at least eight separate harvests.[43]

Site Selection

An ideal site for seaweed growth is anywhere with an annual mean growth rate of $\geq$ four percent per day. It is essential to test the growth rate with lines of cuttings in control test plots placed in different locations within the bay. It is also necessary to ensure the following characteristics of the site:

- That it is not located near a river mouth, so salinity stays between 23 and 38 ppm, and the turbidity of the water is within reasonable limits
- That the species of seaweed already growing in the area are flourishing
- That the water temperature remains between 20 and 32 degrees Celsius throughout the year

- That there is enough current and tidal exchange as seaweeds obtain the necessary nutrients for growth directly from the water. Therefore, this water must be flushed and replaced
- That the site does not exert a negative impact on natural coastal carbon stocks, particularly those associated with seagrass meadows
- That the site minimises negative interactions with other uses of the coastal zone as well as minimises environmental impacts from seaweed farming[44,45]

Case: Seaweed Carbon Farming Study
In 2020, Oceans 2050 announced the launch of the Seaweed Carbon Farming study. The 15-month study will quantify carbon sequestration by seaweed farms in five continents, advancing the scientific basis for seaweed aquaculture as a climate change mitigation solution while contributing to ocean restoration and ultimately creating market incentives to catalyse this solution. The project will involve a global network of leading seaweed scientists and seaweed farm operators sampling sediments from 19 seaweed farms in 12 countries, providing empirical evidence for the carbon sequestration rates. The results will be presented in a peer-reviewed scientific paper. Following the study, Oceans 2050 will develop and submit a voluntary carbon offset market methodology for approval. Once approved, the methodology will enable the issuance of carbon credits by seaweed farmers to global buyers.[46]

Marine Protected Areas

Marine Protected Areas (MPA) and MPA networks were initially conceived as a tool for repairing damage to over-exploited fish stocks and habitats and for conserving biodiversity. They have recently been recognised as essential tools to help protect carbon sinks from habitat loss and environmental degradation so that they can continue to sequester carbon. While all carbon sinks are valuable, MPAs in this context generally focus on protecting coastal wetlands, seagrasses, and mangroves.[47,48,49,50] A set of guidelines is listed below to help MPA and MPA network managers assess the carbon storage capacity of target habitats, identify those areas representing the best and most stable carbon stores, and ensure no net loss of overall stored carbon:

- *Step 1: Identify habitats and species that function as potential carbon sinks*: This requires a coastal and estuarine habitat assessment. The methods used include:
  - Mapping carbon sequestering (carbon sink) ecosystems, habitats, and species using spatial data
  - Measuring the area/size of these ecosystems and habitats
  - Developing carbon-budget models to measure the carbon stored in as many ecosystems, habitats, and species as possible
- *Step 2: Describe the carbon flux system, including carbon sources and the sinks identified in Step 1, and the risks to them*: This involves conducting a risk assessment to identify carbon sinks most vulnerable to climate change and human activities. If protected, restored and/or enhanced, these areas can mitigate climate change. Risk assessments enable the prioritisation of which carbon sinks require immediate protection. The methods used include:
  - Producing an integrated map of the most critical marine and coastal carbon sinks
  - Modelling climate change impacts on the target area
  - Collecting and compiling geospatial socio-economic data on human activities that could affect the marine area
  - Identifying or developing a risk assessment methodology and conducting a risk assessment of the vulnerability of the target area to human activities and climate change
- *Step 3: Determine whether the carbon flux system is vulnerable to impacts of climate change that can be mitigated by MPAs or MPA networks*: Intact marine and coastal carbon sinks reduce carbon in the atmosphere. To ensure the continuity of this function, the most important known carbon sinks may need protection. Efforts should be focused

on ensuring that the priority carbon sink sites are protected and that enough of them are protected (number/size). The methods used include:

- Identifying the carbon sinks within existing MPAs, identifying, and protecting additional areas according to the priorities set out in Step 2, as resources allow
- Ensuring MPA management plans include conservation objectives to protect carbon stores, enabling carbon sequestration to continue within the area, with appropriate management activities to accomplish it
- Establishing a community of experts to promote best practices and develop monitoring and management protocols
- Developing outreach materials about the role of coastal and open-ocean ecosystems, habitats, and species as carbon sinks
- Engaging stakeholders throughout the process to address social and economic considerations

• *Step 4: If impacts on the system from climate change identified in Step 3 can be mitigated by MPAs or MPA networks, topical specialists should estimate the trends and timescale over which the impacts are expected and trigger a re-evaluation of the boundaries of the MPA. Alternatively, they can design the MPA or MPA network to be robust to these changes*: Predictive models and in situ monitoring will increase the understanding of carbon production and storage processes and enable better management programmes to maximise carbon uptake within the MPA network. The methods include:

  - Supporting research on carbon production and storage processes in the target MPAs
  - Using predictive models as well as in situ monitoring
  - Developing a monitoring protocol to monitor the integrity of those ecosystems, habitats, and species that sequester carbon
  - Integrating the carbon system monitoring protocol into the existing MPA network monitoring protocol and activities
  - Carrying out monitoring as feasible

- Refining the carbon-budget models with in situ data as they become available
- Seeking the involvement of coastal communities in monitoring[51]

Case: Estimating Scotland's Blue Carbon Resources in the Inshore Marine Protected Area Network
Scotland has estimated the blue carbon habitats in the Nature Conservation MPAs and Special Areas of Conservation in its inshore waters. It found that blue carbon habitats in the inshore MPA network produce a minimum of 248,000 tonnes of organic carbon and 30,000 tonnes of inorganic carbon per year. The sediment stores in the inshore MPAs accumulate 126,000 tonnes of organic carbon and 348,000 tonnes of inorganic carbon per annum. The result is that integrating the carbon value of these habitats into decisions relating to marine management would improve the protection provided to these habitats and further enhance their capacity to provide a carbon sink.[52]

Ecosystem-Based Adaptation

The traditional approach to protecting people and communities in coastal regions from extreme events has been the use of grey infrastructure, such as seawalls, storm surge barriers, dikes, and levees, for protection from storms and flooding. However, these approaches have led to unintended negative impacts on habitats that can undermine additional flood protection and other services that coastal habitats provide.[53,54,55,56,57] In contrast, EbA harnesses nature's capacity to buffer people and communities against climatic extremes through the sustainable delivery of ecosystem services. EbA focuses on specific ecosystem services that can reduce climatic exposure and involves targeted management, conservation, and restoration activities. At the same time, EbA provides multiple environmental, economic, and social co-benefits to local communities.[58,59,60,61,62] Specific EbA approaches in coastal areas include the following.

Beach Nourishment

Beach nourishment or replenishment is the artificial placement of sand on an eroded shore to maintain the amount of sand present in the foundation of the coast. This process compensates for natural erosion and protects the area from storm surges. Beach nourishment also aims to maintain beaches for recreational activities and tourism. The process involves dredging material (sand, pebbles) from a source area (either offshore or inland) to feed the beach where erosion occurs. Table 9.2 summarises the various beach nourishment techniques available.[63] Sediment used for beach nourishment should be indistinguishable from native site sediment in terms of colour, shape, size, mineralogy, compaction, organic content, and sorting. However, typically fill material does not precisely match the native sediment, so a compatibility analysis is necessary to consider similar fill material. Furthermore, efforts should be taken to avoid and/or mitigate impacts during beach nourishment projects by considering the following:

- *Turbidity*: The amount of turbidity needs to be limited by:
  - Using sediments that closely resembles native sediment
  - Selecting sediments with a low percentage of silt and clay
  - Limiting the rate and total volume of placement
  - Using turbidity curtains to contain suspended solids
  - Conducting fill placement when tidal elevation and wave energy is low
- *Coverage*: Sediment should not be placed in one large area but divided into smaller sections to leave undisturbed areas of existing biological resources. For this approach, future nourishment efforts should alternate which sections receive fill. Nourishment activities should also avoid sensitive habitats and areas with high ecological value. Projects should also avoid productive biological seasons

- *Compaction*: Beach nourishment projects should not increase sand compaction. Compaction affects water retention, permeability, exchanges of gases and nutrients and may reduce nesting success by impeding nest excavation and preventing hatchling emergence
- *Project timing*: The acquisition of fill material and construction phase of any project should be timed to avoid the most productive biological seasons, regular storm seasons, and high beach visitor-use times. Birds, turtles, fish, crabs, and shrimps are just some animals that use the sandy intertidal area. The preferred time for beach nourishment projects depends on the species inhabiting or exploiting the area and the nature and location of the projects

Table 9.2 Beach Nourishment Techniques

Technique	Description
Beach nourishment	• Sand is spread over the beach where erosion is occurring • It compensates for shore erosion and restores the recreational value of the beach • Wind distributes the sand onshore and in dunes
Backshore nourishment	• Sand is stockpiled on the backshore (the area that is only exposed to waves during extreme events) to strengthen the dunes against erosion and breaching during storms • Sand may deplete significantly during storms
Shoreface nourishment	• This is the nourishment of the area between the low water mark and the fair-weather wave base • The reduction of wave energy leads to enhanced accumulation at the beach • Combining this technique with beach nourishment can strengthen the entire coastal profile
Medium-scale (five million cubic metres) nourishment	• This is the nourishment of channel walls or locations where tidal channels erode the coast

(continued)

Table 9.2 (continued)

Technique	Description
Large-scale nourishment	• This has been tested in the Netherlands with the "Sand Motor", which involves the placement of sand in the shoreface and above • The Sand Motor, with a lifespan of around 20 years, acts as a source of sediment supply, with sand redistributed by waves and currents to beaches and dunes over distances of several kilometres • Unlike traditional techniques, the Sand Motor relies on the natural forces of wind and waves rather than mechanical energy

- *Minimising impact on benthic invertebrate populations*: Projects should determine which populations are present and where they occur and use compatible sediment to limit negative impacts to benthic invertebrates. Other ways of reducing impacts on invertebrate survival and recovery include:
  - Conducting beach nourishment projects when invertebrate populations are at their seasonal low on the beach
  - Extending the time between renourishment episodes to permit recovery
  - Placing fill in short sections to leave undisturbed regions for enhanced recovery
  - Limiting the volume and rate of sediment placement to allow motile invertebrates to migrate upwards
  - Maintaining the natural beach profile to preserve habitat area[64,65]

Seagrass Conservation and Restoration

Seagrasses can reduce current velocities, dissipate wave energy, and stabilise sediment, particularly in shallow waters. Reducing the height of waves reaching the shore can decrease wave inundation. Meanwhile, stabilising sediment can help seagrasses accrete with sea-level rise.[66] Seagrass conservation and restoration initiatives aim to induce a change in ecological state, from low structural complexity (usually unvegetated seafloor) to a more complex form (vegetated).[67] There are numerous methods to establish seagrass, including:

- *Plugs*: Plugs of seagrass with the associated sediment can be harvested using a core tube. Core tubes are used to extract plugs from the donor bed and transport them in the tube to the planting site
- *Staple:* Plants are dug up using shovels, the sediment is shaken from the roots and rhizomes in the process, and the plants with the roots and rhizomes are placed in flowing seawater tanks for holding until made into planting units
- *Peat pot*: This method involves a person cutting plugs while another holds out the peat pots and arranges them in a floating tray. As the tray fills up, they may be sunk to the bottom until moved to the planting site. Planting can be done in many ways, including a person loosening the sediment while the other person installs the peat pot in the bottom[68]

There are also a variety of techniques in seed-based restoration:

- *Buoy-deployed seeding (BuDS)*: The BuDS technique involves the collection of mature reproductive shoots that are placed in mesh nets attached to buoys, suspended above plots to be restored with the aim that negatively buoyant seeds, when released, will settle over the desired restorative plot. The collection of reproductive shoots is relatively easy and rapid, and BuDS can be deployed over relatively large spatial scales
- *Dispenser injection seeding*: Seeds are mixed with local sediment to create a sediment-seed mixture for injection into the substrate using modified sealant guns. A predetermined amount of seed is mixed with sieved fine-grained sediment, loaded into sealant tubes, and injected into the sediment with a depth of 1–4 cm

Oyster Reef Conservation and Restoration

Oyster reefs provide a variety of ecosystem services, including shoreline stabilisation and coastal defence and increased resilience of the local economy, for example, by providing habitat for commercially and recreationally important fish and shellfish.[69] Oyster reefs serve as natural coastal buffers, absorbing wave energy directed at shorelines and reducing erosion caused by sea-level rise and storms.[70,71] However, native oyster species of the family Ostreidae that were once dominant worldwide have been decimated globally due to overfishing, eutrophication, and oyster reef degradation.[72]

Usually, an area requiring restoration is either "recruitment limited", "substrate limited", or both and assisted regeneration or reconstruction methods are required. The most common action in oyster restoration projects is deploying oysters, either as spat-on shell or individual larger juvenile seed oysters, to jumpstart restoration efforts to help alleviate recruitment limitation. The aim is to increase broodstock, enhancing initial oyster populations, and increasing the larvae supply to nearby oyster habitats.[73] At times, there may be a limitation on available substrate, requiring the construction of reefs. If the area is substrate- and recruitment-limited, then the constructed reef will need to be seeded with juvenile oysters. There are various factors to consider when selecting material for constructed reefs, as detailed in Table 9.3.[74]

Coral Reef Conservation and Restoration

Coral reefs can attenuate waves, reducing the height of waves reaching the shore and decreasing wave inundation. When corals grow at sea-level rise, this attenuation service can be maintained. In addition to reducing coastal flooding and erosion, coral reef conservation and restoration enhances carbon sequestration and storage, supports ecotourism, enhances recreational opportunities, and provides habitat for recreational and commercial species.[75,76] Nonetheless, coral reefs globally have experienced a decline over the past several decades due to storm

Table 9.3 Factors to Consider When Selecting Oyster Reef Material

Factor	Description/Questions
Recruitment	Will the oysters be set on the selected reef material?
Wave energy	Higher-wave energy areas usually need larger, more durable, heavier reef-building substrate to ensure durability
Benthic characteristics	Heavy reefs may sink in soft mud while shell or other hard bottom substrate can support the weight of the reef
Purpose of the reef project	If the oyster reef is for shoreline erosion protection, it needs to be constructed from suitable materials
Sedimentation	If the reef is constructed in a high sediment accumulation area, it should be designed in a way to withstand those conditions
Sanctuary and public health	Will harvesting be allowed in the area? Does the selected reef material allow for harvest or prevent it?
Conservation status of the restoration site	Do the selected reef material and reef design comply with conservation designations?
Public and regulatory acceptance of the material	Materials considered natural (shell, stone, clay, etc.) could have greater public and regulatory acceptance than others (concrete, plastic, etc.)
User group conflict	Will the reef material interfere with or enhance recreational or commercial fishing? Will it affect boating?
Reef material acquisition and placement	A particular material may be well-suited for the site but locally unavailable. Material costs, transportation costs, and logistics need to be considered

damage, temperature anomalies, and disease, prompting many conservation efforts to protect remaining populations and accelerate recovery trajectories. These efforts include developing regional propagation and restoration programmes.

An emphasis on ecological restoration to reef degradation has increased the popularity of the coral gardening methodology. This involves the outplanting of corals through fragmentation with an intermediate nursery phase to imitate and accelerate reef recovery processes. Specifically, when a portion of coral is broken off, it can be taken to a nursery to stabilise and generate new coral colonies, which can then be placed onto the reef to promote live coral cover on the substrate. Coral nurseries provide a place for coral fragments to survive, propagate, and grow in large quantities, enabling the growth of coral seed from smaller pieces for local reintroduction or degraded areas in neighbouring zones. Nurseries can be either land-based (ex-situ) or ocean-based (in situ), from which individuals are outplanted onto degraded reefs.[77,78]

Land-Based Nurseries

Land-based nurseries are located on the coast, in tanks either out in the open or laboratories. Land-based nurseries' benefits include the frequent monitoring of coral fragments, reduced exposure to diseases given the control of environmental conditions, and optimal growth any time of the year. Land-based nurseries allow for micro fragmentation, a technique that determines the minimum fragmentation sizes of coral colonies, enabling the reproduction of massive coral species that grow slower and can be reared in land-based nurseries. The installation of land-based nurseries requires:

- The installation of water tanks, pumps, valves, pipes, and a drainage system
- Shading, such as the use of screens, to control light and temperature when the nursery is installed outdoors
- Water to be kept at optimal levels for coral growth

Ocean-Based Nurseries

When a site is selected for installing an in situ nursery, various factors need considering, including water quality, shelter, accessibility, and tidal potential. The aim is to place fragments in nurseries with appropriate and similar conditions (depth, water temperature, salinity, sedimentation, etc.) to the outplanting sites to increase the chance of survival. Usually, these nurseries are floating or fixed structures:

- *Floating*: Floating nurseries are helpful when materials are limited, with fragments placed at different depths for their photoacclimation. Floating structures can be attached to marine soil with anchors or placed onto old reef structures, or suspended at mid-water using floats or on the surface
- *Fixed*: Fixed structures are anchored to marine soil and are the best option for shallow-water settings. They consist of modular structures that can hold hundreds or even tens of thousands of fragments[79]

Case: Great Barrier Reef's Reef Restoration and Adaptation Program
In 2018, the Australian Government provided $6 million for the Reef Restoration and Adaptation Program (RRAP) consortium to determine the feasibility of intervening at scale on the Great Barrier Reef (GBR) to help it adapt to, and recover from, the effects of climate change. The interventions needed to protect crucial ecological functions and economic and social values of the GBR, be logistically feasible to deploy at scale and be affordable to deploy across entire reef scapes. Over 18 months, RRAP conducted investigations into small-, medium-, and large-scale reef interventions, involving more than 150 experts from over 20 organisations across the globe. The study found successful intervention was possible and could double the likelihood of sustaining the GBR in good condition by 2050. In 2020, RRAP embarked on a 10-year research and development (R&D) programme to develop, test, and risk-assess novel interventions to help increase the reef's resilience and sustain critical functions and values. The R&D programme is funded through the $100 million allocated for reef restoration and adaptation science as part of the $443.3 million partnership between the Australian Government's Reef Trust and the Great Barrier Reef Foundation. The R&D programme is focusing on interventions that involve:

- Cooling and shading the GBR to help protect it from the impacts of climate change
- Assisting GBR species to adapt to the changing environment
- Supporting natural restoration of damaged and degraded reefs[80]

Mangrove Conservation and Restoration

Mangroves can attenuate waves, reducing wave inundation. They also capture sediment and so help counteract coastal erosion. Mangroves accrete (the vertical build-up of soil) as sea level rises, ensuring these coastal protection services can be maintained. Mangroves can also slow storm surge water flows if mangrove areas are wide enough. The co-benefits of mangrove conservation and restoration include increased resilience of the local economy by providing habitat for commercially and recreationally important fish and shellfish, preserved natural storm defences, and enhanced climate change mitigation from carbon sequestration and storage.[81,82] Nonetheless, vast areas of mangrove forests have been cleared for urban development, industrialisation, agricultural land reclamation, timber and charcoal production, and shrimp farming.[83] Mangrove conservation and restoration can be achieved through natural regeneration or artificial regeneration.

Natural Regeneration

Natural regeneration uses naturally occurring mangrove propagules to restock degraded sites. Specifically, regeneration is from direct, freely falling, and dispersed mangrove propagules, where species composition of the regenerated forest depends on the species types and combinations of the adjacent forest from where the propagules dispersed. The biological and physical factors determining this approach's success include forest conditions, tides, and soil stability. If natural regeneration is viable, stressors that may prevent natural succession from occurring need to be removed to enhance natural recovery. This can be achieved through hydrological restoration (re-modification and re-establishing the ecosystem's original conditions, including tidal regimes).[84]

Artificial Regeneration

Artificial regeneration entails the direct planting of desired propagules and saplings and rarely uses small trees of chosen species at the restoration site. The use of propagules and nursery-raised saplings is the most common method of mangrove restoration. Several advantages of using artificial regeneration include controlling species composition and distribution, introducing genetically improved stocks, and controlling pest infestation. Artificial regeneration may be valuable under a range of conditions, including:

- Planting or sowing may be required when natural supplies of propagules are limited due to lack of nearby parent trees or lack of hydrological connection to these trees (reducing the dispersal of propagules)
- Planting may be done to reintroduce specific valuable species that have been lost from an area
- Plantings provide valuable education and cultural opportunities, creating a lasting commitment and ownership of those involved
- In severely eroded areas, mangrove planting on remaining bunds can provide short-term relief by delaying erosion of those bunds[85,86]

Case: Seacology Supporting the Permanent Protection of Mangroves in Kenya

Lamu Island is situated in the Indian Ocean near Kenya's border with Somalia. A UNESCO World Heritage Site, the island is known for its biodiversity, with mangrove forests, seagrass beds, and coral reefs creating one of the richest fishing areas along the Kenyan coast. The Lamu Archipelago has extensive mangrove forests, covering over 85,000 acres. Mantondoni Village is close to a 1,112-acre mangrove forest threatened by illegal clear-cutting, erosion, and improper waste disposal. Furthermore, people with few economic opportunities cut the trees for firewood and construction materials. The degradation of the mangroves reduces fish populations and diminishes protection from sea-level rise and storms. Seacology supports establishing a Community Forest Association that will work with the Kenyan Forest Service to protect the mangrove forest. Community members will assess the mangrove areas and develop a plan for managing the forest and replanting degraded areas. Also, an information campaign will raise awareness of the importance of mangroves. Furthermore, the community will develop a beekeeping programme as an alternative source of income.[87]

Coastal Wetlands (Non-Mangrove) Conservation and Restoration

The Ramsay Convention defines coastal wetlands as "*areas of marsh, fen, peatland or water, whether natural or artificial, permanent or temporary, with water that is static or flowing, fresh, brackish or salt, including areas of marine water the depth of which at low tide does not exceed six metres*". Non-mangrove coastal wetlands, including saltwater marshes, estuaries, and lagoons, can help attenuate waves and reduce wave inundation. They may also store water during times of high tide, reducing flooding of coastal areas. Coastal wetlands can trap sediment and so may vertically build-up soil as sea level rises. The co-benefits of coastal wetland conservation and restoration include increased flood storage capacity, preservation of natural storm defences, improved water quality, enhanced climate change mitigation through carbon sequestration, supporting of ecotourism through fishing, hunting, and wildlife viewing activities, and the providing of habitat for recreational and commercial species.[88,89,90] However, coastal wetlands are severely threatened, suffering from severe degradation, alteration, or loss due to various activities, including wetland reclamation, pollution, and drainage. They are also considered to be the ecosystem most sensitive to global sea-level rise.[91,92]

The restoration of wetlands should re-establish as far as possible the ecological integrity of the degraded ecosystem, where ecological integrity refers to the condition of the ecosystem, particularly the structure, composition, and natural processes of its biotic communities and physical environment.[93] Methods for coastal wetland restoration can be grouped into the following:

- *Active restoration: This* requires humans to control and intervene regularly to restore, recreate, or improve ecosystem processes in coastal wetlands. This can involve reshaping the topography of the wetlands, rechannelling the water flow through water control facilities, such as division dikes and soil transplantation, and artificially planting vegetation
- *Passive restoration*: This involves eliminating influencing factors that lead to the degradation or destruction of coastal wetlands and restoring degraded coastal wetlands to a healthy state under natural conditions. For example, the improvement of wetland hydrological connectivity and the recovery of underground hydrology can accrete sediment retention and lower soil erosion
- *Creation*: Creation refers to the process of turning lands on sites where no coastal wetlands existed previously into one. This requires appropriate local conditions, particularly suitable hydrology. To reduce losses of tidal marsh, new marshes are often created on sites where no marshes existed previously[94]

Case: Salt Marsh Pilot in the Netherlands

On the far northeast coast of the Netherlands, salt marshes develop naturally with sediment from the port of Delfzijl and the Eems-Dollard estuary. They mitigate the effects of subsidence and sea-level rise. Human-made salt marshes were constructed along the coast of Delfzijl as part of the project Marconi Buitendijks. The project serves as a pilot to develop knowledge of human-made salt marshes. The existing sea walls were relocated and reinforced while a pioneer salt marsh was constructed along the coast, consisting of a pilot salt marsh of 15 hectares, a bird breeding island, and a salt marsh park of 13 hectares. The pilot salt marsh was carried out with three main goals:

1. Creating a natural land–water boundary to improve ecosystem quality
2. Developing knowledge on how to design and construct a pioneer salt marsh at a location that is not suitable yet for salt marsh development
3. Developing knowledge on how the design and construction affect the development of a human-made salt marsh

The pilot salt marsh was intensively monitored between November 2018 and September 2020. The monitoring involved measuring sedimentation and erosion rates, development of tidal creeks, bed level, flooding frequency, vegetation cover, and density at different temporal and spatial scales.[95]

Notes

1. Lindsay Wylie, Ariana E. Sutton-Grier, and Amber Moore, "Keys to Successful Blue Carbon Projects: Lessons Learned from Global Case Studies," *Marine Policy* 65 (2016).
2. FAO and IOC/UNESCO UNEP, "Blue Carbon. A Rapid Response Assessment," (2009), https://www.researchgate.net/profile/Christian_Nellemann/publication/304215852_Blue_carbon_A_UNEP_rapid_response_assessment/links/5769a81308ae2d7145ba854d/Blue-carbon-A-UNEP-rapid-response-assessment.pdf?origin=publication_detail.
3. Jill T. Greiner et al., "Seagrass Restoration Enhances "Blue Carbon" Sequestration in Coastal Waters," *PLOS ONE* 8, no. 8 (2013).
4. Jennifer Howard et al., "The Potential to Integrate Blue Carbon into MPA Design and Management," *Aquatic Conservation: Marine and Freshwater Ecosystems* 27, no. S1 (2017).
5. Inti Carro et al., "Building Capacity on Ecosystem-Based Adaptation Strategy to Cope with Extreme Events and Sea-Level Rise

on the Uruguayan Coast," *International Journal of Climate Change Strategies and Management* 10, no. 4 (2018).
6. Wylie, Sutton-Grier, and Moore, "Keys to Successful Blue Carbon Projects: Lessons Learned from Global Case Studies."
7. UNEP, "Blue Carbon. A Rapid Response Assessment."
8. Greiner et al., "Seagrass Restoration Enhances "Blue Carbon" Sequestration in Coastal Waters."
9. Howard et al., "The Potential to Integrate Blue Carbon into MPA Design and Management."
10. Oliver Moraes, "Blue Carbon in Area-Based Coastal and Marine Management Schemes – a Review," *Journal of the Indian Ocean Region* 15, no. 2 (2019).
11. UNEP, "Blue Carbon. A Rapid Response Assessment."
12. Catherine E. Lovelock et al., "Assessing the Risk of Carbon Dioxide Emissions from Blue Carbon Ecosystems," *Frontiers in Ecology and the Environment* 15, no. 5 (2017).
13. IUCN, "Coastal Blue Carbon Ecosystems: Opportunities for Nationally Determined Contributions. Policy Brief," (2016), https://portals.iucn.org/library/node/48422.
14. Oscar Serrano et al., "Chapter 28—Conservation of Blue Carbon Ecosystems for Climate Change Mitigation and Adaptation," in *Coastal Wetlands*, ed. Gerardo M. E. Perillo, et al. (Elsevier, 2019).
15. AGEDI, "The Abu Dhabi Blue Carbon Demonstration Project: Building Blue Carbon Projects—An Introductory Guide," (2014), https://www.grida.no/publications/326.
16. Moraes, "Blue Carbon in Area-Based Coastal and Marine Management Schemes—A Review."
17. Qingqing Zhao et al., "A Review of Methodologies and Success Indicators for Coastal Wetland Restoration," *Ecological Indicators* 60 (2016).
18. Howard et al., "The Potential to Integrate Blue Carbon into MPA Design and Management."
19. Xiuzhen Li et al., "Coastal Wetland Loss, Consequences, and Challenges for Restoration," *Anthropocene Coasts* 1, no. 1 (2018).
20. Zhao et al., "A Review of Methodologies and Success Indicators for Coastal Wetland Restoration."

21. Julianna J. Renzi, Qiang He, and Brian R. Silliman, “Harnessing Positive Species Interactions to Enhance Coastal Wetland Restoration,” *Frontiers in Ecology and Evolution* 7, no. 131 (2019).
22. Global Nature Fund, “Mangrove Restoration Guide,” (2015), https://www.globalnature.org/bausteine.net/f/8281/GNF_Mangrove_Handbook_2015.pdf%3Ffd%3D0.
23. UNEP-Nairobi Convention/USAID/WIOMSA, “Guidelines on Mangrove Ecosystem Restoration for the Western Indian Ocean Region,” (2020), https://www.nairobiconvention.org/CHM%20Documents/WIOSAP/guidelines/GuidelinesonMangroveRestorationForTheWIO.pdf.
24. Aaron M. Ellison, Alexander J. Felson, and Daniel A. Friess, “Mangrove Rehabilitation and Restoration as Experimental Adaptive Management,” *Frontiers in Marine Science* 7, no. 327 (2020).
25. Howard et al., “The Potential to Integrate Blue Carbon into MPA Design and Management.”
26. AGEDI, “The Abu Dhabi Blue Carbon Demonstration Project: Building Blue Carbon Projects—An Introductory Guide.”
27. Moraes, “Blue Carbon in Area-Based Coastal and Marine Management Schemes—A Review.”
28. Carlos M. Duarte et al., “Assessing the Capacity of Seagrass Meadows for Carbon Burial: Current Limitations and Future Strategies,” *Ocean & Coastal Management* 83 (2013).
29. Carlos M. Duarte, Tomás Sintes, and Núria Marbà, “Assessing the CO_2 Capture Potential of Seagrass Restoration Projects,” *Journal of Applied Ecology* 50, no. 6 (2013).
30. Greiner et al., “Seagrass Restoration Enhances “Blue Carbon” Sequestration in Coastal Waters.”
31. Fernando Tuya et al., “Artificial Seagrass Leaves Shield Transplanted Seagrass Seedlings and Increase Their Survivorship,” *Aquatic Botany* 136 (2017).
32. Richard K. F. Unsworth et al., “Sowing the Seeds of Seagrass Recovery Using Hessian Bags,” *Frontiers in Ecology and Evolution* 7, no. 311 (2019).

33. Diogo Paulo et al., "Open Coast Seagrass Restoration. Can We Do It? Large Scale Seagrass Transplants," *Frontiers in Marine Science* 6, no. 52 (2019).
34. Duarte, Sintes, and Marbà, "Assessing the CO_2 Capture Potential of Seagrass Restoration Projects."
35. M. M. van Katwijk et al., "Guidelines for Seagrass Restoration: Importance of Habitat Selection and Donor Population, Spreading of Risks, and Ecosystem Engineering Effects," *Marine Pollution Bulletin* 58, no. 2 (2009).
36. Marieke M. van Katwijk et al., "Global Analysis of Seagrass Restoration: The Importance of Large-Scale Planting," *Journal of Applied Ecology* 53, no. 2 (2016).
37. Yi Mei Tan et al., "Seagrass Restoration Is Possible: Insights and Lessons from Australia and New Zealand," *Frontiers in Marine Science* 7, no. 617 (2020).
38. Secretariat of the Pacific Regional Environment Programme, "Seagrass Restoration Guidelines for Kiribati," (2020), https://www.pacificclimatechange.net/sites/default/files/documents/Guidelines%20for%20Seagrass%20Restoration%20in%20Kiribati.pdf.
39. The Blue Carbon Initiative, "Blue Carbon Activities," https://www.thebluecarboninitiative.org.
40. Halley E. Froehlich et al., "Blue Growth Potential to Mitigate Climate Change Through Seaweed Offsetting," *Current Biology* 29, no. 18 (2019).
41. Lieve M. L. Laurens, Madeline Lane, and Robert S. Nelson, "Sustainable Seaweed Biotechnology Solutions for Carbon Capture, Composition, and Deconstruction," *Trends in Biotechnology* (2020).
42. Carlos M. Duarte et al., "Can Seaweed Farming Play a Role in Climate Change Mitigation and Adaptation?" *Frontiers in Marine Science* 4, no. 100 (2017).
43. Indian Ocean Commission, "The Farming of Seaweeds," (2012), https://www.researchgate.net/profile/Syed_Jafar2/post/What_are_the_reasons_for_decline_the_yield_of_seaweed_crop/attachment/5a6a0b364cde266d58865689/AS%3A586729060831232%401516898517356/download/a-bl759e.pdf.
44. Ibid.

45. Duarte et al., "The Role of Seaweed Farming in Climate Change Mitigation and Adaptation."
46. Oceans 2050, "Seaweed Project," https://www.oceans2050.com/seaweed.
47. Moraes, "Blue Carbon in Area-Based Coastal and Marine Management Schemes – a Review.".
48. Callum M. Roberts et al., "Marine Reserves Can Mitigate and Promote Adaptation to Climate Change," *Proceedings of the National Academy of Sciences* 114, no. 24 (2017).
49. Howard et al., "The Potential to Integrate Blue Carbon into MPA Design and Management."
50. Roberts et al., "Marine Reserves Can Mitigate and Promote Adaptation to Climate Change."
51. Commission for Environmental Cooperation, "Guide for Planners and Managers to Design Resilient Marine Protected Area Networks in a Changing Climate," (2012), http://www3.cec.org/islandora/en/item/10856-guide-planners-and-managers-design-resilient-marine-protected-area-networks-in.
52. Marine Scotland, "Blue Carbon Resources in Scotland's Inshore Marine Protected Area (MPA) Network," http://marine.gov.scot/information/blue-carbon-resources-scotlands-inshore-marine-protected-area-mpa-network.
53. Richard Munang et al., "Climate Change and Ecosystem-Based Adaptation: A New Pragmatic Approach to Buffering Climate Change Impacts," *Current Opinion in Environmental Sustainability* 5, no. 1 (2013).
54. Carro et al., "Building Capacity on Ecosystem-Based Adaptation Strategy to Cope with Extreme Events and Sea-Level Rise on the Uruguayan Coast."
55. Emily J. Powell et al., "A Review of Coastal Management Approaches to Support the Integration of Ecological and Human Community Planning for Climate Change," *Journal of Coastal Conservation* 23, no. 1 (2019).
56. Rebecca Van Coppenolle and Stijn Temmerman, "Identifying Global Hotspots Where Coastal Wetland Conservation Can

Contribute to Nature-Based Mitigation of Coastal Flood Risks," *Global and Planetary Change* 187 (2020).
57. P. E. T. Edwards, A. E. Sutton-Grier, and G. E. Coyle, "Investing in Nature: Restoring Coastal Habitat Blue Infrastructure and Green Job Creation," *Marine Policy* 38 (2013).
58. Munang et al., "Climate Change and Ecosystem-Based Adaptation: A New Pragmatic Approach to Buffering Climate Change Impacts.".
59. Carro et al., "Building Capacity on Ecosystem-Based Adaptation Strategy to Cope with Extreme Events and Sea-Level Rise on the Uruguayan Coast.".
60. Powell et al., "A Review of Coastal Management Approaches to Support the Integration of Ecological and Human Community Planning for Climate Change.".
61. Van Coppenolle and Temmerman, "Identifying Global Hotspots Where Coastal Wetland Conservation Can Contribute to Nature-Based Mitigation of Coastal Flood Risks."
62. Edwards, Sutton-Grier, and Coyle, "Investing in Nature: Restoring Coastal Habitat Blue Infrastructure and Green Job Creation."
63. ClimateADAPT, "Beach and Shoreface Nourishment," https://climate-adapt.eea.europa.eu/metadata/adaptation-options/beach-and-shoreface-nourishment.
64. National Park Service, "National Park Service Beach Nourishment Guidance," (2012), National Park Service Beach Nourishment Guidance.
65. Jeroen Speybroeck et al., "Beach Nourishment: An Ecologically Sound Coastal Defence Alternative? A Review," *Aquatic Conservation: Marine and Freshwater Ecosystems* 16, no. 4 (2006).
66. UNEP, "Options for Ecosystem-Based Adaptation in Coastal Environments: A Guide for Environmental Managers and Planners," (2016), https://www.unep-wcmc.org/system/dataset_file_fields/files/000/000/380/original/Options_for_Ecosystem_based_Adaptation_in_Coastal_Environments_low-res.pdf?1462462607.
67. Paulo et al., "Large Scale Seagrass Transplants."
68. NOAA/National Centers for Coastal Ocean Science, "Guidelines for the Conservation and Restoration of Seagrasses in the United

States and Adjacent Waters," (1998), https://repository.library.noaa.gov/view/noaa/1672.

69. Powell et al., "A Review of Coastal Management Approaches to Support the Integration of Ecological and Human Community Planning for Climate Change."
70. UNEP, "Options for Ecosystem-Based Adaptation in Coastal Environments: A Guide for Environmental Managers and Planners."
71. Michael W. Beck et al., "Oyster Reefs at Risk and Recommendations for Conservation, Restoration, and Management," *BioScience* 61, no. 2 (2011).
72. Romuald Lipcius et al., "Overcoming Restoration Paradigms: Value of the Historical Record and Metapopulation Dynamics in Native Oyster Restoration," *Frontiers in Marine Science* 2, no. 65 (2015).
73. Lesley P. Baggett et al., "Guidelines for Evaluating Performance of Oyster Habitat Restoration," *Restoration Ecology* 23, no. 6 (2015).
74. Society for Ecological Restoration and The Nature Conservancy, (2019), https://www.natureaustralia.org.au/what-we-do/our-insights/scientific-papers/shellfish-reef-restoration-guidelines/.
75. Powell et al., "A Review of Coastal Management Approaches to Support the Integration of Ecological and Human Community Planning for Climate Change."
76. UNEP, "Options for Ecosystem-Based Adaptation in Coastal Environments: A Guide for Environmental Managers and Planners."
77. D. Hesley et al., "Citizen Science Benefits Coral Reef Restoration Activities," *Journal for Nature Conservation* 40 (2017).
78. United Nations Environment Programme, "Coral Reef Restoration as a Strategy to Improve Ecosystem Services," (2020), https://www.icriforum.org/coralrestoration/
79. MAR Fund, "Training Guide for Coral Reef Restoration," (2020), https://www.icriforum.org/wp-content/uploads/2020/11/Training-Guide-for-Coral-Reef-Restoration.pdf.
80. Reef Restoration and Adaptation Program, "The Program," https://gbrrestoration.org/the-program/.
81. Powell et al., "A Review of Coastal Management Approaches to Support the Integration of Ecological and Human Community Planning for Climate Change."

82. UNEP, "Options for Ecosystem-Based Adaptation in Coastal Environments: A Guide for Environmental Managers and Planners."
83. Babak Kamali and Roslan Hashim, "Mangrove Restoration without Planting," *Ecological Engineering* 37, no. 2 (2011).
84. Ibid.
85. Wetlands International, "Mangrove Restoration: To Plant or Not to Plant?" (2018).
86. UNEP-Nairobi Convention/USAID/WIOMSA, "Guidelines on Mangrove Ecosystem Restoration for the Western Indian Ocean Region."
87. Seacology, "Matondoni Village," https://www.seacology.org/project/matondoni-village/.
88. Powell et al., "A Review of Coastal Management Approaches to Support the Integration of Ecological and Human Community Planning for Climate Change."
89. UNEP, "Options for Ecosystem-Based Adaptation in Coastal Environments: A Guide for Environmental Managers and Planners."
90. Ken Schoutens et al., "Nature-Based Shoreline Protection by Tidal Marsh Plants Depends on Trade-Offs Between Avoidance and Attenuation of Hydrodynamic Forces," *Estuarine, Coastal and Shelf Science* 236 (2020).
91. Charles Cadier et al., "Indicators of Coastal Wetlands Restoration Success: A Systematic Review," *Frontiers in Marine Science* 7, no. 1017 (2020).
92. Renzi, He, and Silliman, "Harnessing Positive Species Interactions."
93. US EPA, "Principles of Wetland Restoration," https://www.epa.gov/wetlands/principles-wetland-restoration.
94. Zhao et al., "A Review of Methodologies and Success Indicators for Coastal Wetland Restoration.".
95. EcoShape, "Salt Marsh Pilot Marconi," (2016), https://www.ecoshape.org/app/uploads/sites/2/2016/07/MarconiSaltmarsh_ReportAnalysisSaltmarshDevelopment_220121_def.pdf.

References

AGEDI. "The Abu Dhabi Blue Carbon Demonstration Project: Building Blue Carbon Projects—An Introductory Guide." (2014). https://www.grida.no/publications/326.

Baggett, Lesley P., Sean P. Powers, Robert D. Brumbaugh, Loren D. Coen, Bryan M. DeAngelis, Jennifer K. Greene, Boze T. Hancock, et al. "Guidelines for Evaluating Performance of Oyster Habitat Restoration." *Restoration Ecology* 23, no. 6 (2015/11/01 2015): 737–45.

Beck, Michael W., Robert D. Brumbaugh, Laura Airoldi, Alvar Carranza, Loren D. Coen, Christine Crawford, Omar Defeo, et al. "Oyster Reefs at Risk and Recommendations for Conservation, Restoration, and Management." *BioScience* 61, no. 2 (2011): 107–16.

Cadier, Charles, Elisa Bayraktarov, Renee Piccolo, and Maria Fernanda Adame. "Indicators of Coastal Wetlands Restoration Success: A Systematic Review." [In English]. *Frontiers in Marine Science* 7, no. 1017 (2020/12/03 2020).

Carro, Inti, Leonardo Seijo, J. Nagy Gustavo, Ximena Lagos, and Ofelia Gutiérrez. "Building Capacity on Ecosystem-Based Adaptation Strategy to Cope with Extreme Events and Sea-Level Rise on the Uruguayan Coast." *International Journal of Climate Change Strategies and Management* 10, no. 4 (2018): 504–22.

ClimateADAPT. "Beach and Shoreface Nourishment." https://climate-adapt.eea.europa.eu/metadata/adaptation-options/beach-and-shoreface-nourishment.

Commission for Environmental Cooperation. "Guide for Planners and Managers to Design Resilient Marine Protected Area Networks in a Changing Climate." (2012). http://www3.cec.org/islandora/en/item/10856-guide-planners-and-managers-design-resilient-marine-protected-area-networks-in.

Duarte, Carlos M., Hilary Kennedy, Núria Marbà, and Iris Hendriks. "Assessing the Capacity of Seagrass Meadows for Carbon Burial: Current Limitations and Future Strategies." *Ocean & Coastal Management* 83 (2013/10/01/ 2013): 32–38.

Duarte, Carlos M., Tomás Sintes, and Núria Marbà. "Assessing the CO_2 Capture Potential of Seagrass Restoration Projects." *Journal of Applied Ecology* 50, no. 6 (2013/12/01 2013): 1341–49.

Duarte, Carlos M., Jiaping Wu, Xi Xiao, Annette Bruhn, and Dorte Krause-Jensen. "Can Seaweed Farming Play a Role in Climate Change Mitigation and Adaptation?" [In English]. *Frontiers in Marine Science* 4, no. 100 (2017/04/12 2017).

EcoShape. "Salt Marsh Pilot Marconi." (2016). https://www.ecoshape.org/app/uploads/sites/2/2016/07/MarconiSaltmarsh_ReportAnalysisSaltmarshDevelopment_220121_def.pdf

Edwards, P. E. T., A. E. Sutton-Grier, and G. E. Coyle. "Investing in Nature: Restoring Coastal Habitat Blue Infrastructure and Green Job Creation." *Marine Policy* 38 (2013/03/01/ 2013): 65–71.

Ellison, Aaron M., Alexander J. Felson, and Daniel A. Friess. "Mangrove Rehabilitation and Restoration as Experimental Adaptive Management." [In English]. *Frontiers in Marine Science* 7, no. 327 (2020/05/15 2020).

Froehlich, Halley E., Jamie C. Afflerbach, Melanie Frazier, and Benjamin S. Halpern. "Blue Growth Potential to Mitigate Climate Change Through Seaweed Offsetting." *Current Biology* 29, no. 18 (2019/09/23/ 2019): 3087–93.e3.

Global Nature Fund. "Mangrove Restoration Guide." (2015). https://www.globalnature.org/bausteine.net/f/8281/GNF_Mangrove_Handbook_2015.pdf%3Ffd%3D0.

Greiner, Jill T., Karen J. McGlathery, John Gunnell, and Brent A. McKee. "Seagrass Restoration Enhances "Blue Carbon" Sequestration in Coastal Waters." *PLOS ONE* 8, no. 8 (2013): e72469.

Hesley, D., D. Burdeno, C. Drury, S. Schopmeyer, and D. Lirman. "Citizen Science Benefits Coral Reef Restoration Activities." *Journal for Nature Conservation* 40 (2017/12/01/ 2017): 94–99.

Howard, Jennifer, Elizabeth McLeod, Sebastian Thomas, Erin Eastwood, Matthew Fox, Lauren Wenzel, and Emily Pidgeon. "The Potential to Integrate Blue Carbon into MPA Design and Management." *Aquatic Conservation: Marine and Freshwater Ecosystems* 27, no. S1 (2017/09/01 2017): 100–15.

Indian Ocean Commission. "The Farming of Seaweeds." (2012). https://www.researchgate.net/profile/Syed_Jafar2/post/What_are_the_reasons_for_decline_the_yield_of_seaweed_crop/attachment/5a6a0b364cde266d58865689/AS%3A586729060831232%401516898517356/download/a-bl759e.pdf.

IUCN. "Coastal Blue Carbon Ecosystems: Opportunities for Nationally Determined Contributions. Policy Brief." (2016). https://portals.iucn.org/library/node/48422.

Kamali, Babak, and Roslan Hashim. "Mangrove Restoration without Planting." *Ecological Engineering* 37, no. 2 (2011/02/01/ 2011): 387–91.

Laurens, Lieve M. L., Madeline Lane, and Robert S. Nelson. "Sustainable Seaweed Biotechnology Solutions for Carbon Capture, Composition, and Deconstruction." *Trends in Biotechnology* (2020/05/06/ 2020).

Li, Xiuzhen, Richard Bellerby, Christopher Craft, and Sarah E. Widney. "Coastal Wetland Loss, Consequences, and Challenges for Restoration." *Anthropocene Coasts* 1, no. 1 (2018/01/01 2018): 1–15.

Lipcius, Romuald, Russell Burke, Danielle McCulloch, Sebastian Schreiber, David Schulte, Rochelle Seitz, and Jian Shen. "Overcoming Restoration Paradigms: Value of the Historical Record and Metapopulation Dynamics in Native Oyster Restoration." [In English]. *Frontiers in Marine Science* 2, no. 65 (2015/09/08 2015).

Lovelock, Catherine E., Trisha Atwood, Jeff Baldock, Carlos M. Duarte, Sharyn Hickey, Paul S. Lavery, Pere Masque, et al. "Assessing the Risk of Carbon Dioxide Emissions from Blue Carbon Ecosystems." *Frontiers in Ecology and the Environment* 15, no. 5 (2017): 257–65.

MAR Fund. "Training Guide for Coral Reef Restoration." (2020). https://www.icriforum.org/wp-content/uploads/2020/11/Training-Guide-for-Coral-Reef-Restoration.pdf.

Marine Scotland. "Blue Carbon Resources in Scotland's Inshore Marine Protected Area (MPA) Network." http://marine.gov.scot/information/blue-carbon-resources-scotlands-inshore-marine-protected-area-mpa-network

Moraes, Oliver. "Blue Carbon in Area-Based Coastal and Marine Management Schemes—A Review." *Journal of the Indian Ocean Region* 15, no. 2 (2019/05/04 2019): 193–212.

Munang, Richard, Ibrahim Thiaw, Keith Alverson, Musonda Mumba, Jian Liu, and Mike Rivington. "Climate Change and Ecosystem-Based Adaptation: A New Pragmatic Approach to Buffering Climate Change Impacts." *Current Opinion in Environmental Sustainability* 5, no. 1 (2013/03/01/ 2013): 67–71.

National Park Service. "National Park Service Beach Nourishment Guidance." (2012). National Park Service Beach Nourishment Guidance.

NOAA/National Centers for Coastal Ocean Science. "Guidelines for the Conservation and Restoration of Seagrasses in the United States and Adjacent Waters." (1998). https://repository.library.noaa.gov/view/noaa/1672.

Oceans 2050. "Seaweed Project." https://www.oceans2050.com/seaweed.

Paulo, Diogo, Alexandra H. Cunha, Joana Boavida, Ester A. Serrão, Emanuel J. Gonçalves, and Mark Fonseca. "Open Coast Seagrass Restoration. Can We Do It? Large Scale Seagrass Transplants." [In English]. *Frontiers in Marine Science* 6, no. 52 (2019/04/07 2019).

Powell, Emily J., Megan C. Tyrrell, Andrew Milliken, John M. Tirpak, and Michelle D. Staudinger. "A Review of Coastal Management Approaches to Support the Integration of Ecological and Human Community Planning for Climate Change." *Journal of Coastal Conservation* 23, no. 1 (2019/02/01 2019): 1–18.

Reef Restoration and Adaptation Program. "The Program." https://gbrrestoration.org/the-program/.

Renzi, Julianna J., Qiang He, and Brian R. Silliman. "Harnessing Positive Species Interactions to Enhance Coastal Wetland Restoration." [In English]. *Frontiers in Ecology and Evolution* 7, no. 131 (2019/04/18 2019).

Roberts, Callum M., Bethan C. O'Leary, Douglas J. McCauley, Philippe Maurice Cury, Carlos M. Duarte, Jane Lubchenco, Daniel Pauly, et al. "Marine Reserves Can Mitigate and Promote Adaptation to Climate Change." *Proceedings of the National Academy of Sciences* 114, no. 24 (2017): 6167–75.

Schoutens, Ken, Maike Heuner, Elmar Fuchs, Vanessa Minden, Tilla Schulte-Ostermann, Jean-Philippe Belliard, Tjeerd J. Bouma, and Stijn Temmerman. "Nature-Based Shoreline Protection by Tidal Marsh Plants Depends on Trade-Offs Between Avoidance and Attenuation of Hydrodynamic Forces." *Estuarine, Coastal and Shelf Science* 236 (2020/05/05/ 2020): 106645.

Seacology. "Matondoni Village." https://www.seacology.org/project/matondoni-village/.

Secretariat of the Pacific Regional Environment Programme. "Seagrass Restoration Guidelines for Kiribati." (2020). https://www.pacificclimatechange.net/sites/default/files/documents/Guidelines%20for%20Seagrass%20Restoration%20in%20Kiribati.pdf.

Serrano, Oscar, Jeffrey J. Kelleway, Catherine Lovelock, and Paul S. Lavery. "Chapter 28—Conservation of Blue Carbon Ecosystems for Climate Change Mitigation and Adaptation." In *Coastal Wetlands*, edited by Gerardo M. E. Perillo, Eric Wolanski, Donald R. Cahoon and Charles S. Hopkinson, 965–96: Elsevier, 2019.

Society for Ecological Restoration and The Nature Conservancy. (2019). https://www.natureaustralia.org.au/what-we-do/our-insights/scientific-papers/shellfish-reef-restoration-guidelines/.

Speybroeck, Jeroen, Dries Bonte, Wouter Courtens, Tom Gheskiere, Patrick Grootaert, Jean-Pierre Maelfait, Mieke Mathys, et al. "Beach Nourishment: An Ecologically Sound Coastal Defence Alternative? A Review." *Aquatic Conservation: Marine and Freshwater Ecosystems* 16, no. 4 (2006/06/01 2006): 419–35.

Tan, Yi Mei, Oliver Dalby, Gary A. Kendrick, John Statton, Elizabeth A. Sinclair, Matthew W. Fraser, Peter I. Macreadie, et al. "Seagrass Restoration Is Possible: Insights and Lessons from Australia and New Zealand." [In English]. *Frontiers in Marine Science* 7, no. 617 (2020/08/14 2020).

The Blue Carbon Initiative. "Blue Carbon Activities." https://www.thebluecarboninitiative.org.

Tuya, Fernando, Francisco Vila, Oscar Bergasa, Maite Zarranz, Fernando Espino, and Rafael R. Robaina. "Artificial Seagrass Leaves Shield Transplanted Seagrass Seedlings and Increase Their Survivorship." *Aquatic Botany* 136 (2017/01/01/ 2017): 31–34.

UNEP-Nairobi Convention/USAID/WIOMSA. "Guidelines on Mangrove Ecosystem Restoration for the Western Indian Ocean Region." (2020). https://www.nairobiconvention.org/CHM%20Documents/WIOSAP/guidelines/GuidelinesonMangroveRestorationForTheWIO.pdf.

UNEP. "Options for Ecosystem-Based Adaptation in Coastal Environments: A Guide for Environmental Managers and Planners." (2016). https://www.unep-wcmc.org/system/dataset_file_fields/files/000/000/380/original/Options_for_Ecosystem_based_Adaptation_in_Coastal_Environments_low-res.pdf?1462462607.

UNEP, FAO and IOC/UNESCO. "Blue Carbon. A Rapid Response Assessment." (2009). https://www.researchgate.net/profile/Christian_Nellemann/publication/304215852_Blue_carbon_A_UNEP_rapid_response_assessment/links/5769a81308ae2d7145ba854d/Blue-carbon-A-UNEP-rapid-response-assessment.pdf?origin=publication_detail.

United Nations Environment Programme. "Coral Reef Restoration as a Strategy to Improve Ecosystem Services." (2020). https://www.icriforum.org/coralrestoration/.

Unsworth, Richard K. F., Chiara M. Bertelli, Leanne C. Cullen-Unsworth, Nicole Esteban, Benjamin L. Jones, Richard Lilley, Christopher Lowe, Hanna K. Nuuttila, and Samuel C. Rees. "Sowing the Seeds of Seagrass Recovery Using Hessian Bags." [In English]. *Frontiers in Ecology and Evolution* 7, no. 311 (2019-August-21 2019).

US EPA. "Principles of Wetland Restoration." https://www.epa.gov/wetlands/principles-wetland-restoration.

Van Coppenolle, Rebecca, and Stijn Temmerman. "Identifying Global Hotspots Where Coastal Wetland Conservation Can Contribute to Nature-Based Mitigation of Coastal Flood Risks." *Global and Planetary Change* 187 (2020/04/01/ 2020): 103125.

van Katwijk, M. M., A. R. Bos, V. N. de Jonge, L. S. A. M. Hanssen, D. C. R. Hermus, and D. J. de Jong. "Guidelines for Seagrass Restoration: Importance of Habitat Selection and Donor Population, Spreading of Risks, and Ecosystem Engineering Effects." *Marine Pollution Bulletin* 58, no. 2 (2009/02/01/ 2009): 179–88.

van Katwijk, Marieke M., Anitra Thorhaug, Núria Marbà, Robert J. Orth, Carlos M. Duarte, Gary A. Kendrick, Inge H. J. Althuizen, et al. "Global Analysis of Seagrass Restoration: The Importance of Large-Scale Planting." *Journal of Applied Ecology* 53, no. 2 (2016/04/01 2016): 567–78.

Wetlands International. "Mangrove Restoration: To Plant or Not to Plant?" (2018).

Wylie, Lindsay, Ariana E. Sutton-Grier, and Amber Moore. "Keys to Successful Blue Carbon Projects: Lessons Learned from Global Case Studies." *Marine Policy* 65 (2016/03/01/ 2016): 76–84.

Zhao, Qingqing, Junhong Bai, Laibin Huang, Binhe Gu, Qiongqiong Lu, and Zhaoqin Gao. "A Review of Methodologies and Success Indicators for Coastal Wetland Restoration." *Ecological Indicators* 60 (2016/01/01/ 2016): 442–52.

10

Blue Financing

Introduction

The ocean sustains the global economy and is a critical source of resources, providing food, jobs, livelihoods, and various ecosystem services necessary for human survival. However, despite recognising the various climatic and non-climatic challenges to the traditional ocean economy, just one percent of its total value has been invested in sustainable projects over the past decade.[1,2] The development of the blue economy is constrained by a lack of fiscal measures, declining development assistance, and limited funding by private investors through foreign direct investments. Also, some countries face sizeable external debt. As such, developing innovative financial approaches is necessary to accelerate the transition to the blue economy.[3,4] This chapter will first provide an overview of the various sources of finance to develop the blue economy before discussing the various financing tools available to facilitate this transition.

R. C. Brears, *Developing the Blue Economy*,
https://doi.org/10.1007/978-3-030-84216-1_10

Sources of Finance for Developing the Blue Economy

There are various sources of finance to develop the blue economy as follows.

Public Financing

Public financing is the capital provided by a national or sub-national government body for goods, services, and infrastructure that serve the public interest. It is the largest category of non-return-seeking capital. In the blue economy, public financing is often allocated towards crucial efforts that do not quickly generate revenue, such as establishing, monitoring, and enforcing marine protected areas (MPAs).[5]

Official Development Assistance

Some countries allocate public funding for official development assistance (ODA). It can be disbursed from one country to another, usually through a national agency (known as bilateral assistance). It can also be disbursed through a multilateral institution, where funding from many nations is pooled. ODA funding can be disbursed as grants towards blue economy projects in developing countries where commercial finance would not consider investing.[6]

Multilateral Development Banks

Multilateral development banks (MDBs) are supranational institutions set up by sovereign states, which are their shareholders. MDBs play a crucial role in delivering sustainable investment in various blue economy initiates, including climate change adaptation and renewable energy projects.[7,8]

Private Financing

There are a few national and international market mechanisms, certification schemes, and foundations that provide sources of finance for implementing blue economy initiatives. For example, voluntary carbon certification schemes generate additional revenue for selling carbon credits from avoided emissions. Other examples include the certification of products from aquaculture and fisheries.[9]

Blended Concessional Finance

Blended concessional finance is an essential tool that MDBs and development finance institutions (collectively known as "DFIs") can use, in cooperation with donors and other development partners, to increase finance for private sector blue economy initiatives. Commercial investors, businesses, and project developers are often constrained by the risk-return profiles associated with investments. Investments with critical public good dimensions may possess good business models and positive projected returns, but associated risk and uncertainty deter commercial investors from providing financing. Furthermore, immature local financial markets, along with information asymmetries and market failures, further discourage private investors. Public support through blended concessional financial approaches helps address these issues by improving the risk-return profile of investments, therefore attracting commercial financing. A set of guiding principles on implementing blended concessional finance transactions are listed in Table 10.1.[10,11]

Case: Asian Development Bank's Oceans Financing Initiative
The Asian Development Bank's (ADB) Oceans Financing Initiative supports ADB developing member countries to catalyse financing for projects that help protect and restore marine ecosystems and promote sustainable blue economies. The initiative leverages public sector funds to create investment opportunities that attract financing from various sources, including the private sector. Various financing tools, including

revenue guarantees and blue bonds, are used to reduce project risks and make them bankable. The initiative aligns with the ADB's Action Plan for Healthy Oceans and Sustainable Blue Economies and aims to expand ADB investment and technical assistance to $5 billion between 2019 and 2024. The initiative's focus areas are:

- *Blue economy*: Creating inclusive livelihoods and business opportunities in sustainable tourism and fisheries
- *Ecosystem management*: Protecting and restoring coastal and marine ecosystems and key rivers
- *Pollution control*: Reducing land-based sources of marine pollution, including plastics, wastewater, and agricultural runoff
- *Sustainable infrastructure*: Improving sustainability in port and coastal infrastructure development[12]

Table 10.1 Guiding Principles for Implementing Blended Concessional Finance Transactions

Principle	Description
1. Additionality/rationale for using blended concessional finance	DFI support of the private sector should contribute beyond what is available or otherwise absent from the market. It should not crowd out the private sector
2. Crowding in and minimum concessionality	DFI support to the private sector should, as much as possible, contribute to catalysing market development and the mobilisation of private sector resources. Concessional finance crowds-in sustainable private sector investments if it provides the missing element in the overall financing that makes private sector projects commercially financeable and if it creates a demonstration effect of commercial replicability

(continued)

Table 10.1 (continued)

Principle	Description
3. Commercial sustainability	DFI support of the private sector and the impact achieved by each operation should aim to be sustainable. DFI support is expected to contribute to the commercial viability of their clients. Operations supported with concessional funds should contribute to the commercial sustainability of the relevant activity or sector and avoid creating permanent dependency on long-term support
4. Reinforcing markets	DFI assistance to the private sector should effectively and efficiently address market failures and minimise the risk of disrupting or distorting markets or crowding out private finance, including new entrants. Concessional finance should not substitute for, or delay, more sustainable commercial or policy interventions
5. Promote high standards	DFI private sector operations should adhere to high standards of conduct in their clients, including corporate governance, environmental impact, social inclusion, transparency, integrity, and disclosure. If a project receiving concessional funds is unnecessary or poorly designed, it will undermine the development or functioning of markets and the private sector

Environmental Taxes, Permits, and Fees

There are a variety of environmental taxes, permits, and fees that can provide direct funding streams for marine conservation and restoration initiatives as well as maintain or enhance protected areas, including the following:

- *Environmental Taxes*: An environmental tax is a tax whose base is a physical unit (or a proxy of a physical unit) of something that has a proven, specific negative impact on the environment, for example, a pollution tax. The main types of environmental taxes are:
  - *Cost-covering charges*: Based on the polluter pays principle, regulatory costs are paid by those being regulated. This type of tax is levied on those making use of the environment to contribute to or cover the cost of monitoring or controlling that use
  - *Incentive taxes*: This is an environmental tax that is levied purely to change environmentally damaging behaviour and without any intention to raise revenues
  - *Fiscal environmental taxes*: A tax that changes, or intends to change, behaviour can also yield significant revenues over and above those required for related environmental regulation
- *Special use permits*: Special uses are activities that provide a benefit to an individual, group, or organisation seeking access to and use of MPA resources. These uses usually require a permit with certain conditions and involve some degree of management and oversight to protect MPA resources. Permits may be granted for one-time events or activities over a period of time
- *Fees*: Fees are payments made for the purpose of privilege, use, or access. For example, they can be collected for specific activities, such as boating, docking, or anchoring/mooring of watercraft or special use of resources, for instance, within an MPA. Fees have the potential for raising revenue to support conservation activities, such as protection, restoration, and education, maintaining or improving an area, or preventing overuse of popular recreational destinations. Examples of fees are:
  - *Entrance/access fees*: Visitors may be charged entrance fees or access fees to support the costs of visitor services or other operating costs of MPAs. Visitors pay when they enter the MPA, use MPA facilities, or participate in various MPA activities. Individuals can be charged on arrival or through a seasonal, annual, or lifetime pass

– *Concession fees*: Companies, known as concessionaires, who provide services within protected areas may be charged fees to conduct their operations. The concession grants the operator exclusivity within a particular area for a particular activity and for a fixed period. Concessions may be renewed to help provide a consistent source of income[13,14,15,16]

Case: Cocos Island National Park Fee
Cocos Island is 340 miles off Costa Rica's southwestern coast. The island is a UNESCO World Heritage Site and designated a Wetland of International Importance under the Ramsar Convention. In 1978 it was designated a Costa Rican national park, encompassing the entire island and protecting its lands and water. Cocos Island is a crucial habitat for large pelagic species, including dolphins, giant manta rays, and at least eight types of sharks. It is considered one of the world's top 10 scuba diving destinations. There is a $25 fee per person and a $10 diving fee, making it $35 per day for divers and $25 for non-divers. Children under 10 are free. Professional filming of the national park carries a $500 fee. Visiting cruise ships are also charged an admission, depending on the size and number of occupants.[17]

Subsidies

Subsidies can be designed to support activities that provide environmental public goods. Subsidies can be direct (such as financial grants or credits that facilitate investments in environmental technologies) or indirect (for example, tax exemptions and the provision of goods below market price). Finally, subsidies can provide direct price support for producers or consumers. The use of subsidies for the development of the blue economy provides numerous benefits, including:

- *Having immediate effectiveness*: Subsidies act immediately as soon as they are provided
- *Providing competitive advantages for enterprises*: When awarded on a national or regional level, subsidies offer enterprises advantages in international competition
- *Supporting innovation at an early stage*: Subsidies reduce the costs of pioneering products and increase knowledge among customers quickly
- *Promoting blue growth*: Subsidies are invaluable for the diffusion of clean technologies and environmental infrastructure
- *Supporting small- and medium-sized enterprises*: Subsidies are advantageous for small- to medium-sized enterprises as they have limited financial capabilities to internally cross-subsidise new products with revenue from well-established products[18,19]

Case: Scotland's Saltire Tidal Energy Challenge Fund
In 2019, the Scottish Government launched a £10 million fund to help the commercial deployment of tidal power. The award of funding took the form of a grant or loan (or another form of repayable assistance), with individual applications able to apply for up to £5 million, which had to be match funded. Match funding could not come from other Scottish public sector grant funding sources. Eligible projects had to meet the following criteria:

- The project had to be related to developing a material/technical innovation to reduce the cost of tidal energy. Applicants had to provide evidence to demonstrate the technical viability and performance of the tidal device to which the application relates
- The proposal had to relate to the capital costs incurred by a tidal energy project. Applications had to provide a clear breakdown of the capital costs of the tidal device to which the application relates
- The proposal had to relate to a project to be deployed in Scottish waters no later than September 2020 and be delivered within clearly defined and manageable timelines

- The proposal had to clearly set out the requirement for, and added value of, Scottish Government support, including why funding from alternative sources is not possible or appropriate[20]

Payments for Ecosystem Services

Payments for Ecosystem Services (PES) are voluntary transactions where a well-defined ecosystem service (or actions likely to secure it) is "bought" from at least one ecosystem service provider by at least one buyer. The payment is conditional on the provision of that ecosystem service. PES schemes can be implemented in marine and coastal settings, which provide a range of ecosystem services, including carbon sequestration, coastal protection, fish nursery, water purification, and marine diversity. The first step in developing a PES scheme is identifying the ecosystem service (s) of interest, the habitats where it is found, and the biological and physical attributes contributing to the ecosystem service provisioning. These ecosystem services include providing goods with existing market value, such as seafood, the management of environmental risks, such as shoreline protection, and the maintenance of the supply chain and base of operations, such as fish nurseries. There is a range of potential providers and buyers in blue economy-related PES schemes, examples of which are listed in Table 10.2.[21,22]

Case: Channel Payments for Ecosystem Services
Channel Payments for Ecosystem Services (CPES) was a cooperation project to encourage farmers to adopt practices that improve catchment water quality in the coastal regions of Southern England and Northern France. It had a budget of EUR 4 million. It ran for 45 months (2017–2020), with fourteen partners working towards improving the water quality of lakes, rivers, and groundwaters by implementing sustainable PES schemes in six pilot catchments. It is estimated that 60–70% of the waters entering the Channel do not reach a Good Ecological Status due

to sedimentation, low oxygen levels, and excess nutrients. Between 30 and 50% of the groundwater bodies in England and France do not meet the chemical quality standards of the Water Framework Directive due to nitrates. Overall, the project provided an experimental framework to test the implementation of PES in different catchment areas, allowing a generalisation of the commercial and sustainable application of the PES concept across varying geographies.[23]

Table 10.2 Potential Provides and Buyers in Blue Economy-related Payments for Ecosystem Services Schemes

Ecosystem service	Examples of voluntary providers (sellers)	Examples of potential buyers or intermediaries
Carbon sequestration	• Various levels of governments • Indigenous/traditional communities • Coastal property owners • Private entities	• Developers, companies, individuals • Governments for meeting emission goals • Carbon offset brokers
Shoreline protection	Same as for carbon sequestration	• Coastal property owners
Biodiversity	Same as for carbon sequestration	• Government agencies and municipalities • Coastal developers • Insurance/reinsurance companies • Natural resources management agencies • Non-profit organisations
Water quality	• Upstream farmers • Upstream municipalities • Indigenous/traditional communities	• Government agencies responsible for public health and safety • Fishing industry • Coastal communities
Fish nursery	• Same as for carbon sequestration • Commercial and artisanal fishers	• Seafood industry, especially buyers, processors, and retailers • Commercial fishermen • Sports fishermen • Dive and snorkel industry

Tradable Permits

Tradable permits establish overall limits on environmental degradation, such as limiting total pollution or a quota on natural resources extracted/removed. The limit or quota is allocated among market participants, who are then free to trade permits (which represent a defined quantity of their individual quota allocation). Tradable permit programmes are either credit programmes or cap-and-trade programmes. Credit trading allows pollution reductions above and beyond baseline legal requirements to be certified as tradable credits. In a cap-and-trade programme, a total resource access limit (the cap) is defined and then allocated among users. Compliance is established by comparing actual use with the assigned firm-specific cap as adjusted by any acquired or sold permits.[24,25] Tradable permit programmes applicable to the blue economy include water quality trading (WQT) and fishing quota.

Water Quality Trading

WQT schemes can be a cost-effective, environmentally sound solution to improving water quality in rivers, lakes, and coastal waters. Generally, WQT involves a party facing high pollutant reduction costs compensating another party to achieve less costly pollutant reduction with the same or greater water quality benefit. The commodity being traded in WQT schemes is either a discharge allowance (from regulated sources) or reduction credits (from non-regulated sources). WQT is most often associated with nutrients (nitrogen and phosphorous).[26,27]

Fishing Quota

Fishing quota, also known as Individual Transferable Quota, provides a share of the fish catch or fishing effort allowed in a fishery to an individual fisher. Fishing quota is usually specific to an individual fish species as part of a fish stock (a distinct population of a species). Usually, a sustainable fish catch (Total Allowable Catch (TAC)) or amount of fishing effort for a species or stock (Total Allowable Effort (TAE)) is established for a fishing season. The fishing quota allocates a portion of that TAC (as a weight of fish) or TAE (as an amount of fishing gear) to their owners of the fishing quota. Often fishing quota is transferable, with quota owners buying or selling it.[28]

Case: The Maryland Water Quality Trading Program

The Maryland Water Quality Trading Program creates a public market for nitrogen, phosphorous, and sediment reductions. The voluntary programme is a collaborative effort between the Maryland Department of the Environment (MDE) and the Maryland Department of Agriculture (MDA) to enhance the restoration and protection of the Chesapeake Bay and local waters by increasing the pace and reducing the cost of implementation efforts. The basis of the programme is:

- *Generating credits*: Credits, where a credit is defined as one pound of pollutant reduction for one year, can be generated for oyster aquaculture as well as for wastewater through the process of:
  1. Meeting baseline requirements for sector-specific practices
  2. Implementing a best management practice
  3. Demonstrating a load reduction below the established baseline
  4. Submitting a Credit Certification and Registration form to MDE
  5. MDE reviewing the documentation and determining certification
  6. Certified credits getting published on Maryland's Trading Registry to be purchased
- *Purchasing credits*: Purchases of Nitrogen, Phosphorus, and Sediment credits are conducted through individual agreements between buyers and sellers. The value of credits will be determined by the market forces of supply and demand, and their value will be determined through negotiations between the buyer and the seller
- *Registry and market*: MDE and MDA are building a comprehensive online portal to manage all the State's WQT data. This tool, called the Chesapeake Bay Nutrient Trading Registry, will be launched soon. Until the tool is in place, access to the WQT information and marketplace will be provided through the:

- *WQT Register*: The WQT Register provides a ledger of certified credits and their status. The ledger also shows where trades have occurred and who is involved.
- *WQT Market Board*: The WQT Market Board is a place for those interested in buying and selling WQT credits. Users can list the watershed, the vintage (year), and other information about the credits they are looking to buy or sell[29]

Biodiversity Offsets

The exploration, construction, production, and decommissioning of infrastructures in marine areas, including renewable energy and aquaculture development, all have temporary and permanent negative impacts on marine ecosystems. Biodiversity offsets are conservation actions intended to compensate for the residual, unavoidable harm to biodiversity caused by projects to ensure no net loss of biodiversity (the counterbalancing of biodiversity losses with biodiversity gains). These gains can be realised through a variety of actions, including:

- Restoration of habitat in another location, such as restoring coral reefs
- Averted loss, which involves the protection of an area, for example, protecting coastal areas from development
- Management to alleviate or avert pressures that would lead to biodiversity losses, for instance, the use of MPAs.

The implementation of biodiversity offsets can take one of three forms:

- *Ad-hoc projects*: These are delivered by the proponent of development causing biodiversity loss
- *Third-party habitat banks*: Biodiversity credits equivalent to meeting offsetting requirements can be purchased or exchanged
- *In-lieu fees*: Financial compensation for biodiversity impacts are pooled for strategic level conservation projects

Meanwhile, Table 10.3 provides a set of principles to ensure biodiversity offsetting is successful.[30,31,32,33,34]

Case: Gladstone Ports Corporation's Biodiversity Offset Strategy for the Western Basin Dredging and Disposal Project

The Biodiversity Offset Strategy (BOS) is a compliance requirement under Queensland, Australia's Environmental Protection and Biodiversity Conservation (EPBC) Act 1999 for approval of the Western Basin Dredging and Disposal Project. Gladstone Ports Corporation has developed the BOS to provide tangible initiatives to avoid potential impacts on the values of the Great Barrier Reef World Heritage Area, National Heritage Place, and EPBC Act listed threatened migratory species. The strategy aims to provide long-term conservation of threatened and migratory species, including their habitats, that may be impacted by the

Table 10.3 Principles to Ensure the Success of Biodiversity Offsetting

Principle	Description
Mitigation hierarchy	Biodiversity offsets should be considered only as a last resort for residual impacts after avoidance and mitigation have been explored
Equivalence	Demonstration of the balance between biodiversity losses and gains is required
Additionality	Biodiversity offsets should not displace existing communities or activities. They should deliver benefits beyond those that would occur in the absence of the offset project
Continuity	Supply of biodiversity through offset projects requires consideration from a temporal and financial perspective: the point at which no net loss of biodiversity is achieved should be matched to the point of impact and that outcomes are delivered for the duration of the impact or in perpetuity
Compliance success	Non-compliance with biodiversity offset requirements is a significant risk of achieving the aim of no net loss. While the legal responsibility for the success of the offset project is with the project proponent or third party delivering the offset, there needs to be oversight and monitoring of implementation by a third party or regulator to ensure compliance with the offsetting requirements

dredging and disposal project activities. The BOS will fund projects for the region and wider bioregion from 2012 to at least 2020, including the following initiatives:

- Acquisition of high-value ecological land to protect from development
- Coral mapping and restoration
- Habitat enhancement and restoration activities
- Stormwater pollution control
- Upper to lower catchment water quality monitoring and improvement of water quality in the Boyne or Calliope Rivers[35]

Blue Bonds

Bonds are the most significant capital markets in the world. A bond is known as a "debt" or "fixed income" security and is created when two parties come together: one seeks to raise money for specific projects (bond issuer, which is a government or similar institution), and one seeks to lend money to earn interest on it (investor). The bond issuer raises capital by selling bonds to investors, who are then owed money by the issuer. The bond will have a specific life span (term), which is usually of five, ten, or fifteen years and return interest at a specific rate (coupon) for the duration of the term. Investors receive their capital back at the end of the term, along with interest payments at specified intervals during the loan, hence the "fixed income" term. A blue bond is a debt security issued to raise capital to finance the blue economy's transition. Blue bonds can finance projects that:

- Contribute to the good governance of the oceans and coastal habitats
- Deliver long-term value to marine and coastal ecosystems
- Reduce carbon emissions
- Strengthen the resilience and livelihoods of people who depend on oceans and their resources
- Restore, protect, and maintain the biodiversity of marine ecosystems
- Reduce stressors to marine ecosystems
- Use nature-based solutions in marine and coastal infrastructure[36,37]

The concept of blue bonds can also be incorporated within existing green bond frameworks, where the proceeds are earmarked for green projects, including renewable energy, pollution prevention and control, terrestrial and aquatic biodiversity conservation, climate change adaptation, and sustainable water management, among others.[38]

Case: Nordic-Baltic Blue Bond
In 2020, Nordic Investment Bank (NIB) issued a new Nordic-Baltic Blue Bond focusing on water investments. The Baltic Sea is threatened by litter, overfishing, unsustainable fishing, and accidental oil spills, with the most significant threat being eutrophication. The five-year SEK 1.5 billion issue will be used to finance projects within water management and protection. Specifically, the proceeds will be allocated to selected wastewater treatment, water pollution prevention, and water-related climate change mitigation projects. The bond was issued under NIB's Environmental Bond Framework, with proceeds from the transaction allocated to a separate account for onward disbursement of loans to the new water-related projects.[39]

Debt for Nature Swaps

A debt for nature swap is an agreement that reduces a country's debt in exchange for a commitment by the debtor government to protect nature. Specifically, in a debt for nature swap, a third party, typically a biodiversity non-governmental organisation (NGO), agrees to raise funds to help the debtor country buy back its debt at a discount. The NGO is motivated by a recognition that the debtor country's coastal and marine territory is rich in biodiversity and has a desire to see it conserved or restored. At the same time, the creditor nations agree to partial repayment of the debt and to write off the remainder. In exchange for the debt reduction, the debtor country agrees to implement conservation and restoration initiatives, for instance, create an MPA in a critical diverse region, and put the savings from the reduced debt service (in local currency) into a conservation trust fund (CTF) to action the initiative. The investment returns on the CTF provide long-term sustainable funding for the costs of management, monitoring, and conservation programmes. Overall, the benefits include:

- The creditor countries avoid the complete loss of the loaned principle
- The NGO and its donors and stakeholders achieve critical conservation objectives
- The debtor country effectively pays off its hard currency debt in local currency[40,41]

Case: Seychelles' Debt for Nature Swap Protecting the World's Oceans

In 2018, Seychelles became the first country to undertake a debt for nature swap to protect the world's oceans. The initiative involved a partnership between The Nature Conservancy's NatureVest, the President and the Environment, Finance, and Foreign Affairs Ministries of Seychelles, and private foundations. The debt restructuring had support from Seychelles' main bilateral creditors, most of which are members of the Paris Club. The debt-restructuring initiative was facilitated by the NGO The Nature Conservancy, which worked with the Ministry of Environment, Energy and Climate Change to create a Marine Spatial Plan for the Exclusive Economic Zone of Seychelles. A locally registered CTF named Seychelles Conservation and Climate Adaptation Trust (SeyCCAT) was established, which purchased and restructured the debt and manages the endowment and the terms of the debt forgiveness agreement. The swap now directs a stream of Seychelles repayments into the SeyCATT to invest in schemes that build a blue economy, including projects related to MPA management, sustainable fisheries, ecosystem rehabilitation, climate change adaptation, and blue economy businesses.[42]

Public–Private Partnerships

Public–private partnerships (PPPs) are long-term agreements between a government entity and a private entity (usually a private company, one or more NGOs, a charitable foundation, or an academic institution). The private entity provides or contributes to the provision of a public service. PPPs are at times mentioned as vehicles for achieving sustainability goals, such as protecting biodiversity, implementing marine renewable energy systems, or restoring coastal ecosystems for climate change adaptation. The private entity receives a revenue stream, which may be from government budget allocations, user charges, or a combination of both. The private entity generally makes an investment in the venture, even if it is limited, for example, to working capital. In addition to budget allocations and/or user charges, the government may make further contributions, such as providing enabling access to land, contributing existing assets, or providing debt or equity finance to cover capital expenditures. At the end of the PPP contract, the associated asset reverts to government ownership. There are a variety of categories of PPPs, including the following below, while Table 10.4 provides a summary of the benefits of PPPs:

- *Management/service contracting or outsourcing*: The government owns and finances the project but appoints a private sector entity to operate it on its behalf. The private sector's remuneration is structured to be a combination of a fixed management fee and a variable success (performance) fee, which will be based on its ability to meet set targets
- *Leases*: The government owns and finances the construction (capital) phases of the project but leases (rents) it out to a private sector entity for a fixed income to government
- *Concessions*: The government owns and finances the construction (capital) phases of the project but grants a private sector entity the permission (concession) to run it
- *Build, Operate and Transfer arrangements*: The private sector entity funds and manages the project, and takes income generated from it until an agreed threshold is reached before the government takes it over

Table 10.4 Benefits of Public–Private Partnerships

Benefit	Description
Improving the quality of service	• The private sector has an incentive to be efficient while the government has experience in regulating • PPP contracts can stimulate innovation through performance indicators and penalties
Improving cost-effectiveness	PPPs allow the public sector to take advantage of private sector innovation, experience, and flexibility
Increasing investment without an onerous burden on public resources	• PPPs bridge the gap between infrastructure needs and government financial capacity • PPPs can provide public goods and services without recourse to taxpayers
Better allocation of risk	Risks can be allocated to the party best able to manage it at a lower cost
Faster implementation	If payments are linked to service delivery, the private sector has an incentive for expeditious completion
Increased investment in technical innovation	PPPs can create incentives to develop new technologies

- *Joint Venture arrangement*: The government and the private sector agree to own the project jointly, invest their funds (as equity), and share profit from operations under the private sector's management[43,44,45,46,47]

Case: Public–Private Partnership for a Marine Protected Area in the Dominican Republic

In the Dominican Republic, Blue Finance, partnering with local NGOs, signed a 10-year agreement with the government to co-manage the largest MPA in the Caribbean. Specifically, the MPA is jointly managed by a non-profit Special Purpose Entity (SPE) through a PPP. The SPE is the creation of a company controlled by a parent company (in this case, local NGOs) which is a distinct legal entity to help keep liabilities, taxation, and regulations related to the project separate from the core business, therefore isolating risk. Blue Finance acts as a project developer, partnering with the government, communities, NGOs, and entrepreneurs/investors to design and implement the SPE arrangements and co-management agreements. The SPE concentrates on two primary income-generating products: enhancing the Under Water (UW) visitor experience and a Marine Life Exhibit centre. The business model is based on generating revenues from statutory visitor fees and innovative tourism activities. The revenues will improve the MPA management efficiency. Meanwhile, the MPA is expected to improve marine biodiversity and generate a sustainable source of food and income for local communities, opportunities for tourism businesses, and protection for coastal properties and beaches. The MPA also contributes to a climate-resilient economy. The SPE is guided by a stakeholder committee, of public and private citizens, with environmental, social, and financial key performance indicators regularly audited. Through the PPP, the MPA is implementing the following activities:

- Supporting the development of the Marine Spatial Plan that regulates the use and guides the environmental and social objectives of the MPA for the next five years. The plan is based on consultations with local stakeholders, reviewed by the Scientific Committee, and approved by the government
- Improving and monitoring the health of the marine habitats, for example, sustainable fishing and tourism, water quality improvements, strengthening reef resilience
- Compliance and enforcement, for example, development of enforcement management plans, effective patrolling, pro-active surveillance by communities and rangers
- Community engagement and livelihood enhancement, for example, awareness campaigns and creation of new income-generating activities, mainly with fishers
- Support to tourism, for example, MPA branding, UW attractions, visitor centre design

- Management and revenue mechanisms. The company generates its income from statutory MPA user fees and the innovative edutainment visitor centre[48,49]

Conservation Trust Funds

The OECD defines CTFs as "*independent legal entities that bridge donors to implementing organisations by providing and sourcing capital specifically for conservation projects*". CTFs provide long-term financing (usually for at least 10 to 15 years) to government agencies, NGOs, or local community organisations for blue economy-related activities, such as developing environmental plans or strategies, managing MPAs or MPA networks, or implementing ocean conservation initiatives. They are capitalised by multiple sources such as multilateral and bilateral contributions and grants from foundations, NGOs, individuals, and revenue-generating activities. CTFs can be structured as:

- *Endowment funds*: Only the interest from the fund is used for projects. The benefits of endowment funds are that they are suitable for projects that require a long-term source of financing, can cover a CTF's basic operation costs, and can be used to leverage additional sources of funding
- *Sinking funds*: Both the interest and principal are used for financing projects until the fund is depleted. The benefits of sinking funds are that they are suitable when large amounts of money are required on a one-time basis, and they are attractive to donors as they like to see the effects of their money being spent
- *Revolving funds*: The fund is replenished regularly through sources such as fees or taxes. The benefits of a revolving fund are that they can last in perpetuity if the source is financially sustainable, cover a CTF's basic operation costs, and connect ecosystem service beneficiaries with providers[50,51,52,53]

Case: Request for Proposals to design a Mediterranean Conservation Trust Fund

The Association for the Sustainable Financing of Mediterranean Marine Protected Areas (M2PA) issued a Request for Proposals (RFP) to design a Mediterranean CTF dedicated to Mediterranean MPAs to sustain their day-to-day management. M2PA received financial support from the Global Environment Facility and Le Fonds Français pour l'Environnement Mondial to design the most appropriate institutional, legal, and financial structure to manage this CTF and to fund pilot MPAs in Mediterranean countries, with an initial focus on projects in Morocco, Tunisia, and Albania. The scope of the RFP included defining the funding allocation and grant delivery mechanism, precisely, the:

1. Granting procedures, the grant award process, including identifying activities and beneficiaries eligibility criteria, selection process, and timeframe
2. The guidelines for grantees, including reporting requirements, eligible costs, and financial obligations
3. Monitoring and reporting requirements.[54]

Corporate Support

Companies can support blue economy conservation and restoration initiatives from their marketing budget, philanthropy budgets, or generally as part of their Corporate Social Responsibility (CSR) work. There are various types of corporate support, including:

- *Financial*: A corporate sponsor could provide financial support in exchange for mentioning the sponsor's name in advertisements or public service announcements or events programmes, newsletters, or press releases
- *Media*: A media sponsor pays for media coverage or secures spaces or airtime on mass media for conservation or restoration efforts. The sponsor is recognised with advertisements containing their official logo, their name, or by association with an outreach campaign brand

- *In-kind*: This involves the donation of goods or services, such as equipment, boats, and vehicles, for marine and coastal conservation and restoration efforts. Corporate sponsors may also provide volunteers for special events or expert assistance
- *Corporate engagement*: A corporation that shares an affinity for a particular marine or coastal conservation or restoration project can be a suitable partner for sponsorships. Corporate engagement can lead to employee volunteer service days, financial donations to blue economy initiatives, and employee donation matching programmes[55]

Case: Adidas' 'Run For The Oceans' Challenge
In 2017, Adidas partnered with Parley for the Oceans to sell shoes made from ocean plastic. Parley Ocean Plastic® is created from upcycled marine plastic waste intercepted from remote islands, beaches, and coastal communities. It is used as a replacement for virgin plastic to make all Adidas x Parley high-performance sportswear. Specifically, after collecting the plastic waste from coastlines, it is baled and sent to Parley supply chain partners. It is then shredded and reworked to become high-performance polyester yarn: Parley Ocean Plastic™. From there, it is used to create Adidas x Parley sportswear. In 2019, Adidas created the Runtastic app for iOS and Android as part of its 'Run For The Oceans' challenge. For every kilometre tracked on the app during the challenge, $1 was donated by Adidas towards fighting marine plastic pollution.[56,57]

Other Financing Approaches

Other financing approaches that can accelerate the transition towards the blue economy and encourage blue growth include:

- *Reef insurance*: Insurance is where a company or organisation provides guaranteed compensation for specified loss or damage in return for consistent payment of a premium. It can be used as a financial tool to minimise risks to coral reefs and generate revenue for reef preservation and/or restoration

- *Voluntary carbon markets*: Voluntary carbon markets deal with the buying and selling of emission reduction credits (offsets) in markets that are not government regulated. The demand for verified carbon credits is driven by voluntary customer demand. Buyers may be the public driven to reduce their carbon footprint or companies taking action to reduce emissions above and beyond their legal obligation to comply with their own CSR. The various types of carbon offsetting include carbon sequestration in coastal wetlands and mangroves and the reduction of carbon emissions from renewable energy
- *Crowdfunding*: Some NGOs and CTFs are using crowdfunding to raise money from individuals. Crowdfunding refers to a campaign or modality of motivating people to make direct contributions through a dedicated technology platform. Crowdfunding is usually successful when targeted towards a highly marketable concept, such as protecting well-known megafauna. Crowdfunding tools include text-to-donate features, QR codes that link to donor websites, and specific web-based platforms. While crowdfunding may not raise significant amounts of funding compared to multilateral or foundation donors, the campaigns aimed at individuals are effective for educating the public on conservation and protection issues, influencing public opinion, and building support for conservation
- *Conservation enterprise incubators*: These encourage conservation actors to flourish within a specific ecosystem by providing economic and technical support that enables conservation enterprises to become successful and competitive. Incubators, also known as accelerators, are programmes that provide technical assistance, development grants, and/or debt or equity financing to assist small commercial ventures in becoming viable or able to seek follow-on funding. Conservation enterprises are commercial ventures that produce both financial and conservation or ecological benefits[58,59]

Case: World's Largest Community-led Mangrove Carbon Conservation Project

Tahiry Honko is the world's largest community-led mangrove carbon conservation project. The project helps mitigate climate change while building community resilience by preserving and restoring mangrove forests in southwest Madagascar at Bay of Assassins within the Locally Managed Marine Area Velondriake. The project promotes locally led conservation, reforestation, and sustainable use of 1,200 hectares of mangroves alongside initiatives to build alternative livelihoods, including sea cucumber, seaweed farming, and mangrove beekeeping. The project aims to provide a long-term income source for the bay residents by selling Plan Vivo certificates generated by avoiding over 1,300 tonnes of carbon dioxide per annum. The project will run on a 20-year crediting period, with an agreed monitoring schedule where staged payments are received annually in return for meeting performance targets, including prevention of ecosystem conversion, improved land-use management, and ecosystem restoration. The sale of carbon credits will also provide a secure revenue flow to support education, dig wells, provide community health services, and contribute to poverty alleviation and biodiversity conservation in the area.[60,61]

Notes

1. Mansi Konar Rashid Sumaila, Ben Hart, and Melissa Walsh, February 25th, 2020, https://www.wri.org/blog/2020/10/7-ways-bridge-blue-finance-gap.
2. Ingrid van Wees, February 25th, 2020, https://www.adb.org/news/op-ed/catching-wave-blue-finance-ingrid-van-wees.
3. Raghu Dharmapuri Tirumala and Piyush Tiwari, "Innovative Financing Mechanism for Blue Economy Projects," *Marine Policy* (2020).
4. Credit Suisse, "Investors and the Blue Economy," (2020), https://www.credit-suisse.com/media/assets/microsite/docs/responsibleinvesting/spread-blue-economy-report.pdf.

5. Friends of Ocean Action, "The Ocean Finance Handbook: Increasing Finance for a Healthy Ocean," (2020).
6. Ibid.
7. IUCN, "Blue Infrastructure Finance: A New Approach, Integrating Nature-Based Solutions for Coastal Resilience," (2020), https://bluenaturalcapital.org/wp2018/wp-content/uploads/2020/03/Blue-Infrastructure-Finance.pdf.
8. R. C. Brears, *Nature-Based Solutions to 21st Century Challenges* (Oxfordshire, UK: Routledge, 2020).
9. Ibid.
10. IFC, "DFI Working Group on Blended Concessional Finance for Private Sector Projects," (2017), https://www.ifc.org/wps/wcm/connect/a8398ed6-55d0-4cc4-95aa-bcbabe39f79f/DFI+Blended+Concessional+Finance+for+Private+Sector+Operations_Summary+R....pdf?MOD=AJPERES&CVID=lYCLe0B.
11. OECD, "Making Blended Finance Work for the Sustainable Development Goals," (2018), https://www.oecd.org/publications/making-blended-finance-work-for-the-sustainable-development-goals-9789264288768-en.htm.
12. ADB, "ADB Oceans Financing Initiative," (2019), https://www.adb.org/sites/default/files/related/145041/Oceans%20Financing%20Initiative.pdf.
13. Marine Protected Areas Federal Advisory Committee, "Protecting Our Marie Treasures: Sustainable Finance Options for U.S. Marine Protected Areas," (2017), https://nmsmarineprotectedareas.blob.core.windows.net/marineprotectedareas-prod/media/archive/fac/products/mpa-fac-external-finance-report.pdf.
14. OECD, "Marine Protected Areas: Economics, Management and Effective Policy Mixes," (2017), https://www.oecd.org/env/marine-protected-areas-9789264276208-en.htm.
15. Brears, *Nature-Based Solutions to 21st Century Challenges*.
16. EEA, "Environmental Taxes: Implementation and Environmental Effectiveness," (1996), https://www.eea.europa.eu/publications/92-9167-000-6/download.
17. CostaRica.com, "Cocos Island National Park," https://www.costarica.com/attractions/cocos-island-national-park.

18. Marine Protected Areas Federal Advisory Committee, "Protecting Our Marie Treasures: Sustainable Finance Options for U.S. Marine Protected Areas."
19. Brears, *Nature-Based Solutions to 21st Century Challenges.*
20. Scottish Government, "Apply for the Saltire Tidal Energy Challenge Fund," https://www.gov.scot/publications/saltire-tidal-energy-challenge-fund/pages/eligibility-criteria/.
21. Winnie W. Y. Lau, "Beyond Carbon: Conceptualizing Payments for Ecosystem Services in Blue Forests on Carbon and Other Marine and Coastal Ecosystem Services," *Ocean & Coastal Management* 83 (2013).
22. Annabelle J. Bladon et al., "Payments for Ecosystem Services in Developing World Fisheries," *Fish and Fisheries* 17, no. 3 (2016).
23. CPES, "The Project," https://www.cpes-interreg.eu/en/cpes-project/the-project.
24. Tom Tietenberg, "The Tradable-Permits Approach to Protecting the Commons: Lessons for Climate Change," *Oxford Review of Economic Policy* 19, no. 3 (2003).
25. The Allen Consulting Group, "Market-Based Approaches to Marine Environmental Regulation," (2006), https://www.mfe.govt.nz/sites/default/files/market-based-approaches-marine-regulation-06.pdf.
26. Suzie Greenhalgh and Mindy Selman, "Comparing Water Quality Trading Programs: What Lessons Are There to Learn?," *The Journal of Regional Analysis and Policy* 42, no. 2 (2012).
27. US EPA, "Water Quality Trading Assessment Handbook: Can Water Quality Trading Advance Your Watershed's Goals?," (2004), https://nepis.epa.gov/Exe/ZyNET.exe/30005XSX.TXT?ZyActionD=ZyDocument&Client=EPA&Index=2000+Thru+2005&Docs=&Query=&Time=&EndTime=&SearchMethod=1&TocRestrict=n&Toc=&TocEntry=&QField=&QFieldYear=&QFieldMonth=&QFieldDay=&IntQFieldOp=0&ExtQFieldOp=0&XmlQuery=&File=D%3A%5Czyfiles%5CIndex%20Data%5C00thru05%5CTxt%5C00000008%5C30005XSX.txt&User=ANONYMOUS&Password=anonymous&SortMethod=h%7C-&MaximumDocuments=1&FuzzyDegree=0&ImageQuality=r75g8/r75g8/x150y150g16/i425&Display=hpfr&DefSeekPage=x&Search

Back=ZyActionL&Back=ZyActionS&BackDesc=Results%20page&MaximumPages=1&ZyEntry=1&SeekPage=x&ZyPURL.
28. Australian Fisheries Management Authority, "What Is Fishing Quota?" https://www.afma.gov.au/what-fishing-quota.
29. Maryland Department of Environment, https://mde.maryland.gov/programs/Water/WQT/Pages/index.aspx.
30. Anne-Charlotte Vaissière et al., "Biodiversity Offsets for Offshore Wind Farm Projects: The Current Situation in Europe," *Marine Policy* 48 (2014).
31. Holly J. Niner et al., "A Global Snapshot of Marine Biodiversity Offsetting Policy," ibid.81 (2017).
32. Céline Jacob et al., "Marine Ecosystem Restoration and Biodiversity Offset," *Ecological Engineering* 120 (2018).
33. Melissa Bos, Robert L. Pressey, and Natalie Stoeckl, "Effective Marine Offsets for the Great Barrier Reef World Heritage Area," *Environmental Science & Policy* 42 (2014).
34. Céline Jacob et al., "Marine Biodiversity Offsets: Pragmatic Approaches toward Better Conservation Outcomes," *Conservation Letters* (2020).
35. Gladstone Ports Corporation, "Biodiversity Offset Strategy," https://www.gpcl.com.au/bos.
36. Stuart Laing, "Blue Economy and Blue Bonds: An Overview of the Concepts and the Pioneering Case of Seychelles," *Seychelles Research Journal* 2, no. 155–64 (2020).
37. BNCFF, "Blue Bonds: Financing Resilience of Coastal Ecosystems," (2019), Blue Bonds: Financing Resilience of Coastal Ecosystems.
38. ICMA, "The Green Bond Principles," (2016), https://www.icmagroup.org/assets/documents/Regulatory/Green-Bonds/Green-Bonds-brochure-150616.pdf.
39. Nordic Investment Bank, "NIB Issues First Nordic–Baltic Blue Bond," https://www.nib.int/who_we_are/news_and_media/news_press_releases/3170/nib_issues_first_nordic-baltic_blue_bond.
40. Brears, *Nature-Based Solutions to 21st Century Challenges*.
41. Wildlife Conservation Society and Conservation Finance Alliance, "Finance Tools for Coral Reef Conservation: A Guide," (2019),

https://www.conservationfinancealliance.org/news/2019/2/28/cfa-publication-finance-tools-for-coral-reef-conservation-a-guide.
42. The Commonwealth Small States Centre of Excellence, "Case Study Debt-for-Nature Finance Swap," (2019), https://seyccat.org/wp-content/uploads/2019/07/SSCOE-Debt-for-Nature-Seychelles-Case-Study-final.pdf.
43. Anis Chowdhury Jomo KS, Krishnan Sharma, and Daniel Platz, "Public-Private Partnerships and the 2030 Agenda for Sustainable Development: Fit for Purpose," (2016), https://www.un.org/esa/desa/papers/2016/wp148_2016.pdf.
44. Marlies Hueskes, Koen Verhoest, and Thomas Block, "Governing Public–Private Partnerships for Sustainability: An Analysis of Procurement and Governance Practices of PPP Infrastructure Projects," *International Journal of Project Management* 35, no. 6 (2017).
45. Living Oceans, "Sustainable Financing Options for a Marine Protected Area Network in British Columbia," (2014).
46. FAO, "Contract Farming and Public–Private Partnerships in Aquaculture: Lessons Learned from East African Countries," (2018), http://www.fao.org/3/CA0134EN/ca0134en.pdf.
47. AU-IBAR, "A Guide for Developing and Implementing Public–Private Partnership Models for Sustainable Fisheries and Aquaculture Development in Africa," (2019), https://www.au-ibar.org/component/jdownloads/finish/5-gi/3432-a-guide-for-developing-and-implementing-public-private-partnership-models-for-sustainable-fisheries-and-aquaculture-development-in-africa.
48. Blended Finance Taskforce, "Blended Finance," https://www.blendedfinance.earth/paying-for-nature/2020/11/16/blue-finance.
49. Blue Finance, "Dominican Republic: Designing, Financing and Implementing the Effective Management for One of the Largest MPAs in the Caribbean," http://blue-finance.org/?p=1740.
50. OECD, "Reframing Financing and Investment for a Sustainable Ocean Economy," (2020).
51. The Nature Conservancy, "Conservation Trust Funds—A Robust Biodiversity Financing Mechanism," (2012), http://caucasus-nature

fund.org/wp-content/uploads/2012/11/CTF_robust_biodiversity_financing_mechanism.pdf.
52. Conservation Finance Alliance, "Rapid Review of Conservation Trust Funds," (2008), https://www.cbd.int/financial/trustfunds/g-rapidassess.pdf.
53. IIED, "A Review of Conservation Trust Funds for Sustainable Marine Resources Management: Conditions for Success," (2014), https://pubs.iied.org/sites/default/files/pdfs/migrate/16574IIED.pdf.
54. Conservation Finance Alliance, "M2PA Request for Proposals for CTF Design," https://www.conservationfinancealliance.org/news/tag/CTF.
55. Marine Protected Areas Federal Advisory Committee, "Protecting Our Marie Treasures: Sustainable Finance Options for U.S. Marine Protected Areas."
56. Adidas, "Parley Ocean Plastic," https://www.adidas.co.nz/sustainability-parley-ocean-plastic.
57. Runtastic, "Run for the Oceans: Join Us to Fight Marine Plastic Pollution," https://www.runtastic.com/blog/en/run-for-the-oceans/
58. Wildlife Conservation Society and Conservation Finance Alliance, "Finance Tools for Coral Reef Conservation: A Guide".
59. IUCN, "Coastal "Blue" Carbon: A Revised Guide to Supporting Coastal Wetland Programs and Projects Using Climate Finance and Other Financial Mechanisms," (2016), https://nicholasinstitute.duke.edu/content/coastal-blue-carbon-revised-guide-supporting-coastal-wetland-programs-and-projects-using.
60. Plan Vivo, "Tahiry Honko," https://www.planvivo.org/tahiry-honko.
61. "Tahiry Honko: Community Mangrove Project, Southwest Madagascar," (2019), https://www.planvivo.org/Handlers/Download.ashx?IDMF=39070798-4292-4797-9ab2-7db77a9afe63.

References

ADB. "ADB Oceans Financing Initiative." (2019). https://www.adb.org/sites/default/files/related/145041/Oceans%20Financing%20Initiative.pdf.

Adidas. "Parley Ocean Plastic." https://www.adidas.co.nz/sustainability-parley-ocean-plastic.

AU-IBAR. "A Guide for Developing and Implementing Public-Private Partnership Models for Sustainable Fisheries and Aquaculture Development in Africa." (2019). https://www.au-ibar.org/component/jdownloads/finish/5-gi/3432-a-guide-for-developing-and-implementing-public-private-partnership-models-for-sustainable-fisheries-and-aquaculture-development-in-africa.

Australian Fisheries Management Authority. "What Is Fishing Quota?" https://www.afma.gov.au/what-fishing-quota

Bladon, Annabelle J., Katherine M. Short, Essam Yassin Mohammed, and E. J. Milner-Gulland. "Payments for Ecosystem Services in Developing World Fisheries." *Fish and Fisheries* 17, no. 3 (2016/09/01 2016): 839–59.

Blended Finance Taskforce. "Blended Finance." https://www.blendedfinance.earth/paying-for-nature/2020/11/16/blue-finance.

Blue Finance. "Dominican Republic: Designing, Financing and Implementing the Effective Management for One of the Largest MPAs in the Caribbean." http://blue-finance.org/?p=1740.

BNCFF. "Blue Bonds: Financing Resilience of Coastal Ecosystems." (2019). Blue Bonds: Financing Resilience of Coastal Ecosystems.

Bos, Melissa, Robert L. Pressey, and Natalie Stoeckl. "Effective Marine Offsets for the Great Barrier Reef World Heritage Area." *Environmental Science & Policy* 42 (2014/10/01/ 2014): 1–15.

Brears, R. C. *Nature-Based Solutions to 21st Century Challenges.* Oxfordshire, UK: Routledge, 2020.

Conservation Finance Alliance. "M2PA Request for Proposals for CTF Design." https://www.conservationfinancealliance.org/news/tag/CTF.

———. "Rapid Review of Conservation Trust Funds." (2008). https://www.cbd.int/financial/trustfunds/g-rapidassess.pdf.

CostaRica.com. "Cocos Island National Park." https://www.costarica.com/attractions/cocos-island-national-park

CPES. "The Project." https://www.cpes-interreg.eu/en/cpes-project/the-project.

Credit Suisse. "Investors and the Blue Economy." (2020). https://www.credit-suisse.com/media/assets/microsite/docs/responsibleinvesting/spread-blue-economy-report.pdf

EEA. "Environmental Taxes: Implementation and Environmental Effectiveness." (1996). https://www.eea.europa.eu/publications/92-9167-000-6/download.

FAO. "Contract Farming and Public–Private Partnerships in Aquaculture: Lessons Learned from East African Countries." (2018). http://www.fao.org/3/CA0134EN/ca0134en.pdf.

Friends of Ocean Action. "The Ocean Finance Handbook: Increasing Finance for a Healthy Ocean." (2020).

Gladstone Ports Corporation. "Biodiversity Offset Strategy." https://www.gpcl.com.au/bos

Hueskes, Marlies, Koen Verhoest, and Thomas Block. "Governing Public–Private Partnerships for Sustainability: An Analysis of Procurement and Governance Practices of PPP Infrastructure Projects." *International Journal of Project Management* 35, no. 6 (2017/08/01/ 2017): 1184–95.

ICMA. "The Green Bond Principles." (2016). https://www.icmagroup.org/assets/documents/Regulatory/Green-Bonds/Green-Bonds-brochure-150616.pdf

IFC. "DFI Working Group on Blended Concessional Finance for Private Sector Projects." (2017). https://www.ifc.org/wps/wcm/connect/a8398ed6-55d0-4cc4-95aa-bcbabe39f79f/DFI+Blended+Concessional+Finance+for+Private+Sector+Operations_Summary+R....pdf?MOD=AJPERES&CVID=lYCLe0B

IIED. "A Review of Conservation Trust Funds for Sustainable Marine Resources Management: Conditions for Success." (2014). https://pubs.iied.org/sites/default/files/pdfs/migrate/16574IIED.pdf.

Ingrid van Wees. "Catching the Wave of Blue Finance—Ingrid Van Wees." Asian Development Bank, 2020.

IUCN. "Blue Infrastructure Finance: A New Approach, Integrating Nature-Based Solutions for Coastal Resilience." (2020). https://bluenaturalcapital.org/wp2018/wp-content/uploads/2020/03/Blue-Infrastructure-Finance.pdf.

———. "Coastal "Blue" Carbon: A Revised Guide to Supporting Coastal Wetland Programs and Projects Using Climate Finance and Other Financial Mechanisms." (2016). https://nicholasinstitute.duke.edu/content/coastal-blue-carbon-revised-guide-supporting-coastal-wetland-programs-and-projects-using.

Jacob, Céline, Jan-Willem Bochove, Suzanne Livingstone, Thomas White, John Pilgrim, and Leon Bennun. "Marine Biodiversity Offsets: Pragmatic Approaches toward Better Conservation Outcomes." *Conservation Letters* (2020).

Jacob, Céline, Anaïs Buffard, Sylvain Pioch, and Sébastien Thorin. "Marine Ecosystem Restoration and Biodiversity Offset." *Ecological Engineering* 120 (2018/09/01/ 2018): 585–94.

Jomo KS, Anis Chowdhury, Krishnan Sharma, and Daniel Platz. "Public-Private Partnerships and the 2030 Agenda for Sustainable Development: Fit for Purpose." (2016). https://www.un.org/esa/desa/papers/2016/wp148_2016.pdf.

Laing, Stuart. "Blue Economy and Blue Bonds: An Overview of the Concepts and the Pioneering Case of Seychelles." *Seychelles Research Journal* 2, no. 155–64 (2020).

Lau, Winnie W. Y. "Beyond Carbon: Conceptualizing Payments for Ecosystem Services in Blue Forests on Carbon and Other Marine and Coastal Ecosystem Services." *Ocean & Coastal Management* 83 (2013/10/01/ 2013): 5–14.

Living Oceans. "Sustainable Financing Options for a Marine Protected Area Network in British Columbia." (2014).

Marine Protected Areas Federal Advisory Committee. "Protecting Our Marie Treasures: Sustainable Finance Options for U.S. Marine Protected Areas." (2017). https://nmsmarineprotectedareas.blob.core.windows.net/marineprotectedareas-prod/media/archive/fac/products/mpa-fac-external-finance-report.pdf

Maryland Department of Environment. https://mde.maryland.gov/programs/Water/WQT/Pages/index.aspx

Niner, Holly J., Ben Milligan, Peter J. S. Jones, and Craig A. Styan. "A Global Snapshot of Marine Biodiversity Offsetting Policy." *Marine Policy* 81 (2017/07/01/ 2017): 368–74.

Nordic Investment Bank. "NIB Issues First Nordic–Baltic Blue Bond." https://www.nib.int/who_we_are/news_and_media/news_press_releases/3170/nib_issues_first_nordic-baltic_blue_bond.

OECD. "Making Blended Finance Work for the Sustainable Development Goals." (2018). https://www.oecd.org/publications/making-blended-finance-work-for-the-sustainable-development-goals-9789264288768-en.htm.

———. "Marine Protected Areas: Economics, Management and Effective Policy Mixes." (2017). https://www.oecd.org/env/marine-protected-areas-9789264276208-en.htm

———. "Reframing Financing and Investment for a Sustainable Ocean Economy." (2020).

Plan Vivo. "Tahiry Honko." https://www.planvivo.org/tahiry-honko.

———. "Tahiry Honko: Community Mangrove Project, Southwest Madagascar." (2019). https://www.planvivo.org/Handlers/Download.ashx?IDMF=39070798-4292-4797-9ab2-7db77a9afe63.

Rashid Sumaila, Mansi Konar, Ben Hart, and Melissa Walsh. "7 Ways to Bridge the Blue Finance Gap." World Resources Institute, 2020.

Runtastic. "Run for the Oceans: Join Us to Fight Marine Plastic Pollution." https://www.runtastic.com/blog/en/run-for-the-oceans/

Scottish Government. "Apply for the Saltire Tidal Energy Challenge Fund." https://www.gov.scot/publications/saltire-tidal-energy-challenge-fund/pages/eligibility-criteria/.

Selman, Suzie Greenhalgh and Mindy. "Comparing Water Quality Trading Programs: What Lessons Are There to Learn? ." *The Journal of Regional Analysis and Policy* 42, no. 2 (2012): 104–25.

The Allen Consulting Group. "Market-Based Approaches to Marine Environmental Regulation." (2006). https://www.mfe.govt.nz/sites/default/files/market-based-approaches-marine-regulation-06.pdf.

The Commonwealth Small States Centre of Excellence. "Case Study Debt-for-Nature Finance Swap." (2019). https://seyccat.org/wp-content/uploads/2019/07/SSCOE-Debt-for-Nature-Seychelles-Case-Study-final.pdf.

The Nature Conservancy. "Conservation Trust Funds—A Robust Biodiversity Financing Mechanism." (2012). http://caucasus-naturefund.org/wp-content/uploads/2012/11/CTF_robust_biodiversity_financing_mechanism.pdf.

Tietenberg, Tom. "The Tradable-Permits Approach to Protecting the Commons: Lessons for Climate Change." *Oxford Review of Economic Policy* 19, no. 3 (2003): 400–19.

Tirumala, Raghu Dharmapuri, and Piyush Tiwari. "Innovative Financing Mechanism for Blue Economy Projects." *Marine Policy* (2020): 104194.

US EPA. "Water Quality Trading Assessment Handbook: Can Water Quality Trading Advance Your Watershed's Goals?" (2004). https://nepis.epa.gov/Exe/ZyNET.exe/30005XSX.TXT?ZyActionD=ZyDocument&Client=EPA&Index=2000+Thru+2005&Docs=&Query=&Time=&EndTime=&SearchMethod=1&TocRestrict=n&Toc=&TocEntry=&QField=&QFieldYear=&QFieldMonth=&QFieldDay=&IntQFieldOp=0&ExtQFieldOp=0&XmlQuery=&File=D%3A%5Czyfiles%5CIndex%20Data%5C00thru05%5CTxt%5C00000008%5C30005XSX.txt&User=ANONYMOUS&Pas

sword=anonymous&SortMethod=h%7C-&MaximumDocuments=1&FuzzyDegree=0&ImageQuality=r75g8/r75g8/x150y150g16/i425&Display=hpfr&DefSeekPage=x&SearchBack=ZyActionL&Back=ZyActionS&BackDesc=Results%20page&MaximumPages=1&ZyEntry=1&SeekPage=x&ZyPURL.

Vaissière, Anne-Charlotte, Harold Levrel, Sylvain Pioch, and Antoine Carlier. "Biodiversity Offsets for Offshore Wind Farm Projects: The Current Situation in Europe." *Marine Policy* 48 (2014/09/01/ 2014): 172–83.

Wildlife Conservation Society and Conservation Finance Alliance. "Finance Tools for Coral Reef Conservation: A Guide." (2019). https://www.conservationfinancealliance.org/news/2019/2/28/cfa-publication-finance-tools-for-coral-reef-conservation-a-guide.

11

Conclusions

Traditionally, the ocean economy is viewed solely as a mechanism for economic growth, which has led to the degradation of coastal and marine ecosystems due to various climatic and non-climatic trends. In the blue economy, the environmental risks of and ecological degradation from economic activity are mitigated or significantly reduced. Therefore, economic activity is in balance with the ocean ecosystem's long-term capacity to support this activity and remain healthy and resilient. Nevertheless, the concept of the blue economy is yet to be mainstreamed for a variety of reasons. For instance, while our scientific knowledge of the oceans is well developed, there is a lack of understanding of how best to develop and implement technologies, investment strategies, and interdisciplinary partnerships that enhance synergies and reduce trade-offs between sectors in the development of a blue economy. There are many cases of initiatives worldwide that ignore blue economy concepts, even when they are obvious and proven to be beneficial to humans and nature. As such, there is a need for innovative policies, technologies, and financing tools that promote collaboration, facilitate the integrated management of resources, foster partnerships between governments and industry, encourage research and development in new technologies, and

R. C. Brears, *Developing the Blue Economy*,
https://doi.org/10.1007/978-3-030-84216-1_11

scale up investments across established and emerging sectors of the blue economy. These sectors include sustainable fishing, sustainable aquaculture, marine biotechnology, marine renewable energy (MRE), coastal water resources management, and blue carbon ecosystems and ecosystem-based adaptation (EbA).

Traditional fisheries management has focused on single species sustainability for commercially valuable species. However, a growing population and increased demand for seafood is resulting in more than half of fisheries being exploited at their maximum level, and many do not have management measures in place to prevent over-exploitation. Various sustainable fisheries practices are available to rebuild global fish stocks, including economic tools that restrict effort and establish access rights to fisheries to ensure fisheries' long-term sustainability. Other practices include implementing technologies to reduce bycatch and creating marine protected areas (MPA) to protect commercial species' critical habitats or life stages. Furthermore, fishery monitoring programmes allow fisheries to reach their full potential for producing food, revenue, and jobs while protecting ocean ecosystems. An ecosystem approach to fisheries (EAF) is required to halt the decline and maintain stocks in the long term. EAF is an integrated management approach across coastal and marine areas and their natural resources that promotes conservation and sustainable use of the whole ecosystem. EAF views fisheries within their broader ecosystem, with management decisions considering the interactions within and between species and between species and their environment. Meanwhile, ecosystem-based fisheries management (EBFM) recognises the physical, biological, economic, and social interactions in fisheries. EBFM aims to sustain healthy marine ecosystems and the fisheries they support by avoiding degradation of ecosystems, minimising the risk of irreversible change to species and ecosystems, obtaining and maintaining long-term socio-economic benefits without compromising the ecosystem, and generating knowledge of ecosystem processes sufficient to understand the likely consequences of human actions.

Rapid population growth and increased demand for fish has resulted in the intensification of aquaculture, increasing waste generation from production systems. The main source of waste from aquaculture includes feed, chemicals, and pathogens. The primary solution for managing the environmental impacts of aquaculture is the management of feed. Adequate feed and feeding systems can effectively reduce waste, such as sieving the feed to remove dust and broken pellets and ensuring the feed is fed effectively to minimise waste. A set of guiding principles are recommended for developing sustainable aquaculture to ensure that communities prosper, livelihoods are improved, income levels are raised, and farmers and women are empowered. The principles include following the precautionary approach to ensure aquaculture plans account for adverse environmental impacts, involving a wide range of stakeholders in the planning process, and involving the public in the decision-making processes. Integrated multi-trophic aquaculture (IMTA) is where uneaten feed and waste of one species are recaptured and converted into feed, fertilisers, and energy for another species. The sustainability of aquaculture increases with IMTA. It allows the creation of more sustainable production systems because wastes of fish/shrimp production are valued as a resource instead of being considered pollution. Organic aquaculture is the farming of aquatic animals and aquatic plants without using antibiotics, chemicals, and fertiliser, preserving the ecosystem and biodiversity. When it comes to spatial planning of aquaculture, the careful selection of farm sites is the first step for ensuring the success and sustainability of aquaculture. Finally, the ecosystem approach to aquaculture is a strategy for integrating aquaculture into the broader ecosystem. It promotes sustainable development, equity, and resilience of interlinked socio-ecological systems.

Marine biotechnology advances can increase food security, improve human health, foster environmental recovery and restoration, and produce renewable energy. Conventional fish vaccines have increased commercial aquaculture production and reduced the use of chemical therapeutics and feed delivered antibiotics. Marine plants, animals, and microbes produce compounds that have the potential as pharmaceuticals to improve human health. A critical aspect of drug development from marine organisms is the permanent availability of sufficient amounts of organisms and compounds without harming the marine environment. If collection from the natural environment cannot be done sustainably, the supply problem can be solved using marine biotechnology, for example, through aquaculture and genetic engineering. Marine organisms are a valuable source of bioactive compounds that provide an unlimited resource for developing new bioactive products. For instance, marine algae are used as sources of food and food ingredients. Marine resources are becoming an increasingly significant source of active ingredients for various cosmetic products. Marine biotechnology is playing an increasingly important role in the protection and management of the marine environment. For example, bioremediation uses natural or genetically manipulated microorganisms to treat oil spills. Microalgae and macroalgae can provide various biofuels, including biogas, biodiesel, ethanol, hydrogen, or algae biomass used for direct combustion. Furthermore, there are various non-energy products obtained from macroalgae, examples of which include food and fertiliser.

The main types of MRE are offshore wind energy and ocean energy, which comprises energy from waves, tidal/sea currents, and thermal and salinity gradients. Offshore wind energy is the most advanced form of MRE in terms of technology development, policy frameworks, and installed capacity. Offshore wind turbines have their foundations in the water (floating) or on the sea bed (fixed-bottom). Ocean wave and tidal current energy are the two types of ocean energy that are expected to contribute significantly to the future supply of energy, with waves constituting a substantial energy resource with very few environmental impacts from the construction of wave energy facilities. Ocean thermal energy conversion uses the temperature difference between warm seawater at the ocean's surface and cold seawater to produce electricity: The warm seawater is used to produce a vapour that drives a turbine. Salient gradient power is the energy created from the difference between two fluids, commonly fresh and saltwater, where a river flows into the sea. Nevertheless, MRE technology can be detrimental to the environment. Usually, developers pursuing MRE must undertake some form of Environmental Impact Assessment (EIA) before deployment. The assessment presents evidence of likely environmental impacts while an environmental monitoring programme provides early warning of potential environmental damage and ensures mitigation measures are being implemented. Finally, marine spatial planning can accurately identify

MRE locations where there are neither conflicts with other activities or severe threats to biodiversity and the marine environment.

Globally, non-point source pollution has contributed to eutrophication in estuarine and coastal waters, resulting in reduced water quality, loss of habitat and natural resources, and hypoxia. Sea-level rise has led to seawater intrusion into surface water and coastal aquifers, diminishing freshwater resources for human use. Simultaneously, rising demand for water has led to many locations worldwide implementing desalination projects, resulting in various environmental impacts offshore. Furthermore, waterways transport significant amounts of plastic pollution into the oceans, degrading marine ecosystems, and impacting human health. Best management practices (BMPs) can mitigate pollution of surface and groundwater and the ocean in various contexts. Agricultural BMPs are tools that farmers can use to reduce soil erosion and fertiliser runoff, protect water quality on their farms, and effectively manage animal waste while achieving positive environmental outcomes, including protecting the marine environment. In urban settings, BMPs retain runoff locally and allow water to penetrate the ground where the pollutants can be attached to soil particles and degraded in the vegetation's root zone. Forestry BMPs minimise water pollution from forestry operations by addressing sediment and sediment transport, which is the primary source of pollution from silviculture. Watershed management plans can be developed to protect coastal and marine ecosystems from various pollution sources, with plans involving watershed monitoring, stakeholder engagement, and education and outreach initiatives. Aquifer Storage and Recovery (ASR) is used worldwide to provide seasonal water storage, reduce groundwater overdraft, replenish depleted aquifers, and improve the drinking water supply quality. In coastal areas, ASR systems can act as a hydraulic barrier against saltwater intrusion. Regarding desalination projects, EIAs provide the ability for authorities to mitigate and control the environmental impacts of these projects. Finally, to mitigate marine plastic pollution, a framework can be implemented at any level—at the municipal, regional, national, or transnational level—with the last step of the framework establishing what needs to change and the interventions required to enact change.

Blue carbon strategies focus on preserving and enhancing the organic carbon stocks and organic carbon burial capacity of salt marshes, mangroves, and seagrasses, particularly within their soil. The conservation and creation of blue carbon ecosystems also provide numerous benefits, including protection from storms, prevention of coastal flooding and shoreline erosion, regulation of water quality, provision of habitat for commercially important fisheries, and food security for coastal communities. Seaweed production can also help reduce the emissions from agriculture by substituting synthetic fertiliser and lowering methane emissions from cattle when included in cattle feed. MPAs and MPA networks were initially conceived as a tool for repairing damage to overexploited fish stocks and habitats and for conserving biodiversity. They have recently been recognised as essential tools to help protect carbon sinks so that they can continue to sequester carbon. EbA harnesses nature's capacity to buffer people and communities against climatic extremes through the sustainable delivery of ecosystem services. EbA focuses on specific ecosystem services that can reduce climatic exposure and involves targeted management, conservation, and restoration activities. At the same time, EbA provides multiple environmental, economic, and social benefits. Specific EbA approaches in coastal and marine areas include beach nourishment and seagrass, oyster reef, and coral reef conservation and restoration. Additionally, mangrove and non-mangrove coastal wetlands can be conserved and restored to reduce flooding of coastal areas.

The development of the blue economy is constrained by a lack of fiscal measures, declining development assistance, and limited funding by private investors through foreign direct investments. Also, some countries face sizeable external debt. As such, developing innovative financial approaches is necessary to accelerate the transition to the blue economy. There are various sources of finance to develop the blue economy that is low-carbon, efficient, and clean and supports the carrying capacity of the oceans in meeting socio-economic development goals, including public finance, official development assistance, multilateral development banks, private finance, and blended concessional finance. From these sources, there are various financing tools available to facilitate the transition to a blue economy. Environmental taxes, permits, and fees can provide direct funding streams for marine conservation and restoration initiatives and maintain or enhance protected areas. Payments for Ecosystem Services schemes can be implemented in marine and coastal settings to provide a range of ecosystem services, including carbon sequestration, coastal protection, fish nursery, water purification, and marine diversity. Tradable permit programmes, including water quality trading and tradable fishing quotas, can limit marine and coastal environmental degradation. Biodiversity offsets can involve the proponent of developments causing biodiversity loss to implement conservation or restoration projects. Blue bonds can finance blue economy projects that contribute to the good governance of the oceans and coastal habitats and strengthen the resilience and livelihoods of people who depend on oceans and their resources. A debt for nature swap involves a third-party raising funds to help a debtor country buy back its debt at a discount. The creditor nations agree to partial repayment of the debt and writing off the remainder in exchange for the debtor country implementing conservation and restoration initiatives. Public–private partnerships can be developed to implement a range of blue economy projects, such as protecting biodiversity, implementing MRE systems, or restoring coastal ecosystems for climate change adaptation. Conservation trust funds provide long-term financing for blue economy-related activities, such as developing environmental plans or strategies, managing MPAs or MPA networks, or implementing ocean conservation initiatives. Companies can support blue economy conservation and restoration initiatives from their marketing budget, philanthropy budgets, or generally as part of their Corporate Social Responsibility work. Other financing approaches that can accelerate the transition towards the blue economy and encourage blue growth include reef insurance, voluntary carbon markets, crowdfunding, and conservation enterprise incubators.

Index

R. C. Brears, *Developing the Blue Economy*,
https://doi.org/10.1007/978-3-030-84216-1

The manufacturer's authorised representative in the EU is Springer Nature Customer Service Centre GmbH, Europaplatz 3, 69115 Heidelberg, Germany. If you have any concerns regarding our products, please contact ProductSafety@springernature.com

Printed and bound by CPI Group (UK) Ltd, Croydon, CR0 4YY
22/09/2026
02221063-0006